AF564714

Tissue Culture in Forestry

Tissue Culture in Forestry

Dr. Sachin Kumar

Tissue Culture in Forestry

ISBN 978-93-5111-220-4

Published in 2014 in India by

RANDOM PUBLICATIONS

4376-A/4B, Gali Murari Lal, Ansari Road
New Delhi-110 002
Phone : +91-11-43580356, +91-11-23289044
e-mail: randomexports@gmail.com, sales@randompublications.com, info@randompublications.com

Reprinted 2022

Type Setting by : Keystoneprintads, Delhi-110051
Printed at : Replika Press Pvt. Ltd.

Preface

Tissue culture has been proposed as a method for the large-scale clonal propagation of forest species. Plant tissue culture relies on the fact that many plant cells have the ability to regenerate a whole plant (totipotency). Single cells, plant cells without cell walls (protoplasts), pieces of leaves, or (less commonly) roots can often be used to generate a new plant in culture media given the required nutrients and plant hormones. The main advantage of this technique is the production of exact copies of the parent plant.

Plant tissue culture technique offers an excellent opportunity for mass propagation of plants in laboratory test tubes, which are transferred to the field. Besides crop plants, the technique is also applied to regenerate saplings for plantation and regeneration of dwindling forests. Some rare and nearly extinct plant species can be rescued and propagated by this technique.

Due to rapid deforestation and depletion of genetic stocks, concerted efforts must be made to evolve new methods for mass propagation and production of short duration trees with a rapid turnover of biomass and induction of genetic variability for the production of novel fruit and forest trees which are high yielding, resistant to pest and disease associated with increased photosynthetic efficiency. This required genetic manipulation to evolve vigorous and fast growing trees with a short reproductive cycle which can be mass propagated.

It is envisaged that the technology of tissue culture is competent to meet this challenge. Tissue culture techniques have already revolutionized the mass scale propagation of many horticultural crops and several commercial laboratories have been set up in many parts of world for mass production of elite, cloned plant material. However, its exploitation of forest tree species has started only recently.

Tissue culture plantlets have poor photosynthesis efficiency and lack the proper mechanism to control water loss. They need to be hardened gradually by moving them along a humidity gradient in the greenhouse. Once these plants are in the research fields, they are evaluated under field conditions and the data is collected every 6 months. A large number of tissue culture plants that have grown into trees are remarkably uniform and show an increase in biomass production over the conventionally raised plants.

Some of the chapters will provide the reader with a broad review of recent developments and progress in the field in the field of tissue culture.

I thank all members of my team who have helped in the preparation of the book. My special thanks go to "Random Publications" who have published the book.

– ***Dr. Sachin Kumar***

Contents

Preface .... *v-vi*

1. **Ggenetic Modification of Forest Trees** **1**
   - Global System on Conservation and Sustainable Use of Plant Genetic Resources 5
   - Useful Traits by Genetic Transformation 8
2. **Benefits of Urban Forests and Trees** **18**
   - Recreational use of Green Areas 19
   - Urban Growing Conditions 21
   - Urban Characteristics/Environment Analysed 23
   - Trees in the Urban Environment 24
3. **Forest Tree Improvement and Breeding** **27**
   - Understanding the Tree Improvement 30
   - Definition of Trees Outside the Forest 33
4. **Genomics and Forest Biology** **37**
   - Sequenzcing the Poplar Genome 37
   - The Great Diversity Found in Forest Trees 42
   - Proposal to Sequence Populus Genome 45
   - Random Forest for Gene Selection and Microarray Data Classification 48
   - Gene Selection and Classification of Microarray Data Using Random Forest 52
   - Emerging Model Systems in Plant Biology 61
   - Genetic Approach of Stem Secondary Growth 68
5. **Application of Plant Tissue Culture to Agriculture and Forestry** ... **71**
   - Tissue Culture 74
   - Plant Tissue Culture 87

Plant and Forest Biotechnology 91
Applications in Horticulture and Forestry 94

6. Mmodern Position of Tree Development in Forestry 97

Elements of a Genetic Improvement Programme 97
Tree Improvement in Developed Countries 112
Uses of Forest Tree Species in Some Sadc Countries 116
The Status of Tree Improvement 120
Applications in Forest Tree Improvement 131
Effects of Ozone on Forest Trees 134
Methodologies Used in Assessing Ozone Effects 146

7. Iin Vitro Propagation of Forestry 152

Cryopreservation and in Vitro Storage 152
In Vitro Selection 155
Somaclonal Variation 163
Biotechnologies and Tree Improvement 168
Metabolites in In Vitro Regenerated Plantlets 174

8. Forest Trees from Biotechnology 178

Changes Outside the Forest Sector 184
Recent and Future Innovations and Advances 190
The Transition to Plantation Forestry 196
Driving Forces and Outlook to 2050 201

9. Clones in Nature and Foresry 220

Evolution of Technology for Cloning Eucalyptus in Large Scale 232

10. Ggenetically Engineered Trees in Forestry 246

Introduction 246
The Specific Concerns of Certification Bodies 253
Evaluating the Risks 270
Cultural Aspects of Forests 274
Trees, Woods, Culture and Imagination 278

Index 284

1

Genetic Modification of Forest Trees

GMOs are defined as organisms that have been modified by the application of recombinant DNA technology (where DNA from one organism is transferred to another organism). The term "transgenic trees" is also used for GM trees, where a foreign gene (a transgene) is incorporated into the tree genome. One of the first reported trials with GM forest trees was initiated in Belgium in 1988 using poplars. A study carried out in 1999 indicated that, since then, there have been more than 100 reported trials, involving at least 24 tree species-most of which are timber-producing species. The majority of the field trials were carried out in the USA and Canada. Whereas it is estimated that roughly 40 million hectares of transgenic agricultural crops were grown commercially in 1999 (ISAAA figures), there is no reported commercial-scale production of transgenic trees. Information on field trials of GM trees has been published by the OECD and the World Wide Fund for Nature (1999). Traits for which genetic modification can realistically be contemplated in the near future include insect and virus resistance, herbicide tolerance and lignin content.

However, insertion of any gene into a tree species with expected functional results will be a substantial undertaking and insertion of enough genes to confer *e.g.* long-term insect resistance in a perennial species even more so. Virus and insect resistance, in particular, are of major significance for crop plants. By contrast, these traits are not the most important in breeding programmes of forest tree species (poplars being an exception). Reduction of lignin is a valuable objective for species producing pulp for the paper industry; work on this aspect is underway in aspen. A major technical factor limiting the application of genetic modification to forest trees, is the current low level of knowledge regarding the molecular control of traits which are of most interest, notably those relating to growth and stem and wood quality. Genetic modification of these traits remains a distant prospect. Investments in genetic modification technologies should be weighed against the possibilities of exploiting the large amounts of genetic variation, which are generally untapped, available within any single species in nature. Biosafety aspects of

GM trees need careful consideration because of the long generation time of trees, their important role in ecosystem functioning and the potential for long distance dispersal of pollen and seed.

FORESTRY IN DEVELOPING COUNTRIES

Forests cover approximately 30 percent of the world's total land area. They are the source of vital commodities, including raw materials and food and are essential for maintaining agricultural productivity and the environmental well-being of the planet as a whole. They protect soil and water and buffer the effects of wind and rain, thus helping to decrease soil erosion and they are an important sink for carbon dioxide. Forests are also among the most important repositories of biological diversity. Roughly 500 million rural people live in, or close to, forests. Most communities use a variety of forest products, particularly those in developing countries. Plant stems, tubers and fruits provide additional food during hungry seasons or when crops fail; wild animals are harvested for meat and hides; and the forests provide fuelwood, fodder for livestock, medicines and other products and services. The most important trend in forestry in developing countries is the progressive reduction in the area of forests due to changes in land use. Another important trend, evident at a global level, is increasing forest degradation through unmanaged use. When forests are degraded, their productive functions and their capacity as regulators of the environment are reduced, increasing flood and erosion hazards, reducing soil fertility and contributing to the loss of forest products and overall loss of biological diversity.

While forests are being lost, there is growing demand both for environmental services and for wood and wood products which they provide. A forecast by FAO predicts that wood demand is expected to increase by 25 percent from 1996 to 2010. This demand will, increasingly, have to be met by forest plantations, and with decreasing land areas available for forestry, plantation methods will have to be increasingly intensive. This will necessitate better tree improvement programmes in which biotechnology may play a role.

THE WEALTH OF INTRAPOPULATION GENETIC VARIATION

Within populations, four forces (selection, migration, mutation and population size) determine the structure of genetic variation. The interaction of all four factors can be highly complex and their effects are determined by the mating system and the kinds of gene effects that exist. A problem in analysing even this level of genetic evolution is that the kinds of gene effects that exist depend on the effects that other forces of evolution have had on gene actions themselves. To consider how the forces can affect evolution, we can first start with each of them and then consider how their interactions can change predictions. For selection to have an effect on speciation and the structure of forests, there must be a difference among genotypes in the traits affected by survival or reproduction events and the individuals must accordingly leave more or fewer progeny than other genotypes. The genotype is defined as the individuals with a distinctive ensemble of alleles at its gene loci that distinguishes it from other genotypes. This is a multiple gene locus definition, which is often simplified by considering every one of the loci to be independent in the occurrence of its alleles in both frequency and effect.

In this case, we can consider the genotype to be the summation of all independent loci and hence can examine gene effects and dynamics one locus at a time. This is, of course, an oversimplified view since genes are in fact linked on chromosomes and therefore have linkages that correlate their occurrences, and most genes affect processes that are combined in their net effect on traits; furthermore, most genes affect enzymatic pathways that ultimately affect several traits when measured at the level of whole organism performance. These effects are called linkage, epistasis and pleiotropy, respectively, and tie the whole organism together into an integrated set of actions and functions that can survive and reproduce. While undoubtedly true in general, there is little evidence that individuals are so bound by such constrained genetic constitutions and physiological limitations that substantial independence among loci exists. The evidence for the existence of strongly selected, tightly linked 'coadapted gene complexes' is not strong and hence the model of independence can be usefully employed, even though it is surely wrong at least in detail.

Within a gene locus, many alleles may exist and the pairwise interactions among them are called the dominance relations. For a diploid organism, if both alleles at a locus are the same, the individual is said to be a homozygote; if the two alleles are different, the individual is said to be a heterozygote. If the heterozygote has a phenotype intermediate between the two homozygotes, the dominance effect is considered to be zero and the gene action is said to be additive. If the heterozygote is other than intermediate, it may be less than the lesser of the two homozygotes. If its phenotype lies between the homozygotes but is not exactly intermediate and if it appears to be greater than the greater of the homozygotes, the gene action is said to be underdominant, partially dominant or overdominant, respectively.

The effect of selection would depend on the kind of gene action, since directional selection for greater or lesser phenotypes would favour different homozygote and heterozygote states. Selection may also be stabilizing if it favours an intermediate phenotype or disruptive if it disfavours the intermediate and its effect on the frequency of the alleles would depend on the gene actions. Directional selection experiments seem to indicate that most alleles have partial dominance actions, though many may be involved in multiple locus interactions. Since selection would be expected to favour survival and reproduction of certain genotypes, we expect that it would increase the population level of fitness. Selection is therefore usually considered to have a positive effect on the evolutionary potential of a population or species in small increments every generation but may not generate an optimal condition at any one time. For alleles that have partial dominance effects, the dynamics of the rates of displacement of one allele by another are such that the frequency of the allele itself affects its own rate of displacement. When alleles are at either low or high frequency, their rates of change are slower than when they are at an intermediate frequency. Except in small populations where accidental loss can occur, selection would not generally lose alleles that are favoured for adaptation.

Another force of evolution is mutation, which is usually considered to be a random change in gene structure and function that occurs indiscriminately throughout the genome. Since most species have evolved a genetic and physiological level that is close to a sufficient capacity for continued evolution, novel changes are not expected to improve performance or fitness but rather to generally decrease adaptability. However, there are always going to be mutations and while most may be disfavoured by selection, there is likely to be a balance between the rate at which new mutants are introduced and the rate of their elimination. Most mutations occur at low rates for any one locus in any one generation and would not be expected to affect the adaptability of a population unless they are accumulated in small populations. For most conifer species where the frequency of deleterious mutants has been estimated, there are a larger number of lethal alleles carried at one or another locus in most individual trees than in most other plant or animal species. This might indicate that there are many effective loci in trees that can mutate to a harmful state, that the mutation rate is high or that the mutants are favoured in some conditions or are of some advantage in heterozygous conditions. However, independent mutations and selection can generate differences among populations, which can increase fitness in each. Hybridization among populations can then generate high levels of genetic variance in traits that may then be useful in new environments.

A third force of evolution is migration, which can occur among many different kinds of population structures. The input of pollen or seed from other populations that have different allele frequencies may bring in alleles that might otherwise be lost, and if the migration rate is high enough among all populations may induce all populations to have approximately the same frequencies. If the immigrant alleles are favoured, then further immigration would advance the rate of its increase; if disfavoured, immigration would maintain alleles at higher frequencies than expected in a manner similar to mutation; and if neither, would tend to make frequencies homogeneous among all populations. In the presence of divergent selection among populations that exist in different environments, selection would act to eliminate migrant alleles and would maintain population divergence. The fourth force is sampling error induced by small population sizes, which tends to increase the probability that low-frequency alleles are lost by chance. If populations remain at low sizes for several generations, the chances of random loss are accumulated; even populations of more than 20-50 adults can lose alleles that may have started with moderate frequencies.

Since most forest trees seem to carry many deleterious mutants, even if at low frequency at any one locus, another problem is that small populations will suffer from inbreeding depression and may eventually suffer such debilitation from several mildly depressive mutants that the population goes extinct. Individual populations would have to be large enough to sustain viability, and to avoid environmental accidents and catastrophes if no migrants were recruited. There are two problems faced in persistently small populations, the loss of alleles and inbreeding depression; hence, larger population sizes are generally better for avoiding extinction. However, the effects of selection, migration and mutation are confounded with the sampling errors incurred in real populations of finite size. Thus, if populations are separated and selected for different environmental optima, migration can introduce alleles in high frequency that depress fitness. On the other hand, migration can maintain alleles in populations where they might be useful, but would be lost due to sampling accidents in small populations.

It is reasonable to expect that all species are under the influence of several or all of the forces of evolution simultaneously and that the balance between the several forces is not the same for all genes in all populations. Furthermore, since the physical and biotic environments of forests are rarely at an equilibrium, the genetic and ecological dynamics of most species would have to be considered to be in other than a stable state. The actual distribution of the multiple functional forms of alleles at each of the several tens of thousands of gene loci in each of the individuals of a population must therefore be the resultant of the mixture of forces felt at each locus, subject to historical events. The vast possibilities of mixed distributions are overwhelming, although it is also obvious that for most of the species we can study, species are not random mixtures of an infinite number of possible combinations.

GLOBAL SYSTEM ON CONSERVATION AND SUSTAINABLE USE OF PLANT GENETIC RESOURCES

The salient points related to the progress of the global system on conservation and sustainable use of plant genetic resources for food and agriculture which emerged in the seventh regular session of the FAO Commission on Genetic Resources for Food and Agriculture (CGRFA) at Rome during 15-23 May, 1997 may be reviewed as follows:

- A multi-year work programme on agricultural biodiversity established as per decision III/11 of COP shall be pursued by the FAO Commission and CBD Secretariat though cooperation. Similarly, FAO's monitoring/reporting on agricultural biodiversity shall be consistent with CBD, other relevant inter-governmental bodies and world food summit' Plan of Action.

- CBD through its decision II/5 has requested GEF to give priority support to efforts for conservation and sustainable use of agro-biodiversity. Many countries stressed the need to have separate funding! FAO commission requested FAO to provide assistance and guidance to countries on request in implementation/monitoring of the Global Plan of Action (GPA). Further, the monitoring of GPA and updating/reporting on the state of world's PGR were held complementary activities which do not need separate reports (*i.e.*, need only one report each) by countries.
- No changes in the CCGCT (Code of Conduct for Germplasm Collecting and Transfer) or any further work on Code of Conduct for Biotechnology were recommended, till the revision of IUPGR. However, a decision guide on regeneration of accessions in seed collections, prepared by IPGRI, has been made available to the FAO commission for comments by members.
- The existing arrangements between FAO and the 12 CGIAR centres shall be extended, pending the revision of IUPGR.
- Statutes of the inter-governmental technical working group on AGR (Animal Genetic Resources) and PGR were presented. Both working groups have 5 members each for Asia. The AGR were specifically addressed in view of the broadened mandate of the Commission and to facilitate integrated approach to agro-biodiversity.
- The summary statement presented by IPGRI informed the Commission that 0.5 million accession were now covered under the FAO-CGIAR agreements with the designation of 50,000 new accession under this arrangement.
- The summary statement presented by the CBD Secretariat showed that at least three decisions of the COP-III, Buenos Aires, Nov., 1996, were of direct relevance to the ongoing negotiation for revision of IUPGR, namely, Decision III/11-Conservation and Sustainable use of agro-biodiversity, Decision III/15-Access to Genetic Resources, and Decision III/17-Intellectual Property Rights. The COP-III noted the narrow options for legal status of revised International Undertaking, as being, i) a voluntary agreement, ii) a binding instrument, and iii) protocol to CBD. It encouraged parties to implement GPA at national levels. Pollinators and soil-micrograms in agriculture needed special attention (as bio-indicators) and comprehensive reporting by countries to Subsidiary Body on Scientific, Technical and Technological Advice (SBSTTA)/Conference of Parties (COP) of the CBD.
- An identified group of competent people needs to resolve the PGR issues and help prioritise various PGR activities. There is need to identify crops/regions on priority.

- The dialogue between the public and private sector should continue and there is a need to create awareness about PGR.
- An information bank should be established to enhance literacy, management and utilisation of genetic wealth. Emphasis should also be given to regional languages in developing the information base.

World Information and Early Warning System

The FAO's initiative to develop a systematic information base on PGR and related technologies under the World Information and Early Warning System on Plant Genetic Resources (WIEWS) is aimed to collect, collate, maintain and disseminate facts and figures on country profiles, national PGR programmes, *ex situ* collections, database on PGR, existing crop varieties, seed production and supplying agencies, *in situ* on farm conservation, bioprospecting and exchange of seed. The nature of such an activity requires a comprehensive information input and a grid approach for analysis and interpretation of data. The system is open and responsive to queries from all stakeholders and interest groups. In practical terms, it would be relatively simpler to attend to details of the resource availability and the custodian(s) but, considering the importance of implementing CBD and WTO provisions and the imminent revision of the International Undertaking on Plant Genetic Resources, the access to resources for the purpose of replenishment could pose certain difficulties. The access would have to be required from any one of the following sources:

- The *ex situ* seed collections;
- The seed multiplication and supplying agencies; and
- The local communities/community gene banks.

The implications, in each case, could be several and manifold. The access from *ex situ* CG system collections is being debated wherein two clear-cut options have been put forth. The pre-CBD collections are still available freely which means their replenishment would be only a matter of time actually required for the multiplication of their seed. However, in case of requirement of post CBD collections, several access-related issues shall have to be sorted prior to the multiplication of seed for replenishment. The requirement of seed from commercial/public seed multiplication and supplying agencies could be easily met but such a replenishment would not serve the purpose effectively. This being so, the agencies could provide seed of few dominant varieties, which they are currently handling, in contrast with the requirement of a wider genetic base. The replenishment of seed from farmers' varieties/land races faces two different situations:

- The issue of consent/access; and
- The issue of compensation or sharing of benefit accrued from the use of their heritage seed.

The setting up of community seed banks and the Global Biodiversity Fund have definitely been provided as suitable institutions which could help in build up at an appropriate system and come to the rescue in cases of crisis. The setting up of a separate Global Biodiversity Fund is still being debated as also the Farmers' Rights. The development of the *sui generis* systems to meet the countries' own (local) needs, has to be accelerated. Through this mechanism, the whole idea is to encourage local level replenishment through procurement of required seed from nearly/similar climate logical areas.

Some Queries Seeking Answers

Increased investments on *ex situ* conservation are unavoidable with each increment in the quantum of work and management skills. The cost/benefit ratio does not work in such cases because the conservation is for posterity. In this endeavour, the obvious questions seeking answers would be:

- What should be the estimate of funds for genetic resources conservation at the CG Centres, at the National genebanks and at community seed banks?
- What kind of activities we need to take up in the beginning which ensure fast replenishment in cases of crisis?
- What shall be the mechanism of funding-whether the Global Biodiversity Fund or private sector contribution shall be forthcoming or not?
- Which sites for *in situ*/on-farm conservation to begin with?
- How to identify genuine NGOs for their involvement in PGR Conservation?
- What could be a standard format of collecting and collating information for WIEWS on PGR *ex situ* conservation, country status, on-farm conservation and wild relatives? Do the existing proforma suffice?
- What would be a suitable indicator or an appropriate method for monitoring of genetic erosion for use under Early Warning Systems?
- What should be the suitable proforma for multicrop passport data?
- How to generate compatible country reports providing information for the State of the World's PGR and the action on GPA?
- How to minimise diversity and redundancy in use of computer software for various Database Management Activities?

USEFUL TRAITS BY GENETIC TRANSFORMATION

NUTRITIONAL IMPROVEMENT

In addition to its basic calorific value, wheat with its high protein content is an important source of plant protein in the human diet. One of the prominent targets for the application of transgene technology for the nutritional improvement of wheat is targeted at enhancing the grain quality by, i) increasing the protein content, ii) increasing essential amino acids such as lysine, iii) increasing the high molecular weight (HMW) glutenins to improve breadmaking properties of wheat flour, iv) modifying starch composition and v) producing pharma-and neutraceuticals.

Amongst the cereals, the flour of bread wheat, Triticum aestivum, has a superior capability of forming leavened bread. This superiority stems from the structure and composition of its seed storage proteins, which upon hydration can interact to form gluten, an insoluble, but highly hydrated, visco-elastic aggregate that endows the wheat dough with its unique properties. Although the majority of wheat seed storage proteins participate in gluten formation, biochemical and genetic evidence has demonstrated that high molecular weight glutenin subunit (HMW-GS) plays a major role in determining the visco-elastic properties thereby determining bread making qualities. The HMW glutenins are necessary to create a strong dough, which is essential for making high quality, yeast-raised breads. Strong dough traps tiny bubbles of carbon dioxide gas formed naturally by yeast during mixing and subsequent raising thereby enabling the dough to rise forming leavened breads. Dough strength and the ability to contain gas bubbles is known as visco-elasticity and is an important characteristic of wheat with respect to its end product quality. Increasing understanding of the molecular basis of dough visco-elasticity would thus help in developing strategies for minimizing the effect of unfavourable environmental factors on wheat end-use quality.

Genetic transformation of wheat is a key component in a scheme proposing a complete set of approaches to apply biotechnology to improve wheat quality via direct manipulation of HMW-glutenin genes. To alter the amount and composition of these proteins by genetic engineering, a gene encoding a novel, hybrid-subunit of HMW-GS under the control of native HMW-GS regulatory sequences was inserted into wheat using the biolistic approach (Blechl and Anderson, 1996). The HMW-GS 1Ax1 gene which is known to be associated with superior bread making quality, was introduced into the cultivar (Bobwhite) lacking this gene. The introduced 1Ax1 gene under the control of HMW-GS promoter was expressed at high levels and stability was maintained for several generations.

The results demonstrate the feasibility of manipulating the composition of wheat kernels by genetic engineering and have successfully made changes in both the levels and the types of seed storage proteins. This work also demonstrated the usefulness of tissue-specific promoters for the expression of transgenic proteins in the endosperm tissue of wheat. The prospect of achieving the elusive goal of nutritional improvement of wheat brightened further with the improvement in functional properties of wheat dough due to transformation of wheat with high molecular weight subunit genes. Transformation with one or two subunit genes results in stepwise increase in dough elasticity. The expression of a recombinant protein with x-and y-type HMW-GS subunit in transgenic wheat has resulted in altered gluten polymer assembly and composition. This study thus made it feasible to change the glutenin composition by expressing modified HMW-GS in transgenic plants.

The quality improvement studies were extended to durum wheat by He *et al.* 1999 with the introduction of one of the two subunits of HMW-GS subunit genes (1Ax1 or 1Dx5). Analysis of the expression of the additional subunits in the T2 generation using a mixograph, indicated an increase of dough strength and stability. This study demonstrated the feasibility of manipulating durum wheat for superior bread and pasta making qualities. Similarly, the overexpression of the HMW subunit 1Dx5, in transgenic wheat (T-aestivum) resulted in a four-fold increase in this proportion of component in the seed protein and also a corresponding increase in the proportions of the total HMW proteins and glutenins.

However, the overexpression of the 1Dx5 gene was found to be associated with a dramatic increase in dough strength by making it too strong, and, therefore, unsuitable for use in conventional breadmaking. Recently, Alvarez *et al.* (2000) introduced the HMW-GS genes 1Ax1 and 1Dx5 into a commercial cultivar of T-aestivum that already expresses five subunits. The overexpression of 1Dx5 gene increases the contribution of the HMW-GS to a level of 22% of the total protein content. In near future, molecular approaches including genetic transformation and marker-assisted selection will provide an opportunity for improving further the wheat processing qualities.

Modification of wheat starch is presently targeted in various laboratories to improve its potential utility. Wheat grain is predominantly composed of starch, which is a mixture of two polymers, the almost linear amylose molecules, and the heavily branched amylopectin molecules. The ratio of amylose to amylopectin in starch determines its physico-chemical characteristics and thereby its end-use. Wheat flour, low in amylose content is desirable for noodlemaking as it improves noodle texture. In wheat, scientists are thus working towards increasing the amylopectin content of starch thereby reducing amylose content leading to the formation of a value added, low amylose flour. The starch branching enzymes catalyse formation of 1, 6-linkages in the glucan polymer and control the amount of amylopectin produced. In recent years, efforts are underway towards detailed characterization of the starch branching enzymes and starch synthases. This increasing information and characterization of the various components of starch biosynthesis will enable researchers to make rational design of novel starches and alteration in starch levels in other crop plants as well.

Wheat is widely used as an animal feed also for non-ruminants in several developed countries of the world. The phytase of Aspergillus niger is used as a supplement in animal feeds to improve the digestibility and also to improve the bioavailabilty of phosphate and minerals. The phyA gene from Aspergillus niger, encoding for the phytase enzyme has been successfully expressed in transgenic wheat lines by the microprojectile bombardment of immature embryos. The constitutively overexpressed phytase was found to accumulate at high levels in the endosperm. This work may thus open newer avenues for a wider applicability of wheat. Wheat is also an ideal system for the production of novel compounds due to its excellent storage properties and the existence of an efficient processing industry. The potential products include the high value, low-volume compounds such as biologically active proteins and peptides, as well as, high volume, low-cost raw materials for packaging and building. The production of recombinant antibodies in rice and wheat was recently reported with the expression of a medically important, single chain Fv recombinant antibody against the carcino-embryonic antigen.. The recombinant antibody was targeted to the plant cell apoplast and endoplasmic reticulum and was detected in the leaves and seeds of wheat. More significantly, the recombinant antibodies remained active after prolonged seed storage at room temperature thus opening avenues for the further exploitation of wheat for producing high-value, novel compounds.

ENGINEERING NUCLEAR MALE STERILITY

The production of hybrids is an essential component of crop breeding programmes but till date, hybrid wheat has remained elusive! The development of a suitable hybridisation system for wheat requires a high degree of male sterility in all parts of the female parent to avoid self-fertilization. De Block and coworkers have developed a nuclear male sterile system in wheat by introducing the barnase gene under the control of a tapetum specific promoter, the expression of which prevents normal pollen development at specific stages of anther development. This system employed the ribonuclease-inhibitor barstar gene to restore the fertility of male sterile plants. To avoid complicated gene integration patterns, the target tissues were incubated on niacinamide containing medium before bombardment. The authors suggest that the enzyme poly (ADP-robose) polymerase (PARP), which plays a key role in the processes of cell division and recombination, is inhibited by niacinamide, thereby resulting in simple integration pattern of the transgene. Expressing the barnase gene at specific stages of anther development destroys the tapetum, thereby preventing normal pollen development and causes pollen sterility.

RESISTANCE TO BIOTIC STRESS

The integration of transgenic approaches with classical breeding techniques offers a potential chemical-free and environment-friendly solution for controlling pests and pathogens. Wheat is attacked by a number of viral, bacterial and fungal pathogens and also by insect and nematode pests. Introgression of genes from the wild relatives exhibiting resistance to pests and pathogens has been successfully utilized over the years for generation of resistant varieties in wheat. With the development of plant transformation techniques, newer avenues for creating disease resistant and insect resistant crops have been created. Coupled with this is the availability of novel transgenes encoding highly potential anti-microbial peptides, defense-related proteins and enzymes for the production of anti-microbial compounds in crop plants, have greatly enhancing the possibility of engineering crop plants for resistance to pests and pathogens. Fungal pathogens of wheat cause severe crop damages by infecting the spikes, leaves and roots. Amongst the various strategies to introduce fungal resistance by transgenic approach, strengthening the host plant defense by genetic manipulation hold tremendous potential. Biochemical and structural responses against fungal attack include reinforcement of plant cell wall, accumulation of phytoalexins with microbial toxicity, ribosome-inactivating proteins (RIP) that inhibit protein synthesis, antimicrobial peptides and synthesis of other PR proteins,. Genetic engineering allows the expression of foreign genes from distant unrelated species as well as the modification of the usual pattern of expression of an already present gene.

Most of the works on genetic engineering of wheat for resistance against biotic stress have focussed on developing protection against fungal pathogens. Introduction genes encoding for chitinases from barley resulted in increased resistance against Erysiphe graminis. Bliffeld *et al.* (1999) reported the adverse effect of a ribosome inactivating protein on plant regeneration and development. Nonetheless, moderate protection against Erysiphe graminis was reported with the transformation of wheat with a gene encoding for a ribosome inactivating protein from barley. The genes encoding for thaumatin like protein (TLP) and stilbene synthase in transgenic wheat have been shown to improve resistance of T1, T2 progeny plants against the fungal pathogens. An increase in endogenous resistance against Tilletia tritici was achieved with the introduction of virally encoded antifungal protein.

For the engineering of resistance against different pests and pathogens in wheat, genes encoding for viral coat proteins, antifungal proteins, and proteinase inhibitors have been successfully introduced. Most of the introduced genes confer increased resistance to the corresponding pests and pathogens in the transgenic plants. Introduction of canditate genes (lectin, proteinase inhibitor) for insect resistance into wheat have resulted in growth inhibition of insects on transgenic seeds, thereby decreasing the fecundity of insect population.

RESISTANCE TO ABIOTIC STRESS

Traditional approaches at transferring resistance to crop plants are limited by the complexity of stress tolerance traits, as most of these are quantitatively linked traits (QTLs). Nonetheless, the direct introduction of a small number of genes by genetic engineering offers convenient alternative and a rapid approach for the improvement of stress tolerance. Although, present engineering strategies rely on the transfer of one or several genes that encode either biochemical pathways or endpoints of signalling pathways, these gene products provide some protection, either directly, or indirectly, against environmental stresses. Drought is a major abiotic factor that limits crop productivity, thereby causing enormous loss. The genes encoding the late embryogenesis proteins (LEA) which accumulate during seed desiccation, and in vegetative tissues when plants experience water deficiencies have recently emerged as attractive candidates for engineering of drought tolerance. Transgenic approach has been used for successfully introducing and overexpressing the barley HVA1 gene encoding for a late embryogenesis abundant (LEA) protein by Sivamani *et al.* (2000) into wheat by particle bombardment.

Most of the transgenic lines tested displayed improvement in important agronomic traits, including total dry mass and water use efficiency, shoot dry weight, root fresh and dry weights, when plants are grown under soil water deficit conditions. In general, this investigation showed that the transgenic lines expressing the HVA1 gene had improved growth characteristics including an enhanced biomass yield under water deficit conditions. The discovery of novel genes, determination of their expression pattern in response to abiotic stress and an improved understanding of their roles in stress adaptation (obtained by the use of functional genomics) will provide the basis of effective engineering strategies leading to greater stress tolerance (Cushman and Bohnert, 2000).

TRANSGENE SILENCING IN WHEAT

Success at developing improved wheat cultivars through genetic engineering depends on stable and predictable expression of the inserted gene. However, gene silencing is a common phenomenon in the production of transgenic plants and needs to be effectively controlled for the desired results. This is all the more important because gene silencing is an important phenomenon involved during natural plant defense. Gene silencing is a complicated phenomenon as it includes both transcriptional gene inactivation and post transcriptional gene inactivation. The complex and the large genome size of wheat is expected to be prone to silencing by introduced transgenes.

With the development of transformation methodologies for wheat, more information has started to accumulate regarding the inheritance and expression of the introduced transgenes. Information regarding the long term stability of transgenes is thus of immense significance for the use of genetic manipulation as a tool in wheat crop improvement and is necessary for the introduction and subsequent expression of desirable agronomic trait.

DNA methylation plays a significant role in establishing and maintaining an inactive state of the gene by rendering the chromatin structure inaccessible to the transcription machinery. Methylation of DNA is expected to result in reduced gene expression. Muller *et al.* (1996) studied the variability of transgene expression in clonal cell lines of wheat and found a negative correlation between the degree of methylation and marker gene expression. PEG-mediated approach was used to transform protoplasts isolated from suspension cultures of T-aestivum. The integration and expression of the selectable marker gene nptII fused with different promoters, namely, alcohol dehydrogenase, shrunken from maize, and actin1 from rice, was confirmed by Southern analysis and enzyme activity test. Transgenic cell lines under selection were 'protoplasted' and clonal callus lines were cultivated from genetically identical single cells without selection pressure. A reduction/loss of marker gene expression was observed due to a reduction in the nptII transcript level and was seen to be associated with hypermethylation of the integrated DNA. The silencing effect was reversed by a 4-week culture phase on media supplemented with demethylation agent, 5-azacytidine, thus confirming the role of methylation in transgene silencing Demeke *et al.* (1999) studied the inheritance and stability of an Act1D-uidA: nptII expression cassette in T4 and T5 transgenic plants. Based on the histochemical localization of GUS activity, this study demonstrated the lack of any cytoplasmic effect on the inheritance of the transgene. The transgenes maintained a multiple integration pattern similar to that observed in the T1 generation. The transgenic plants which produced low gus and nptII activity in seeds had an intact expression cassette. Southern blot analysis of genomic DNA performed with the methylation sensitive enzyme, HpaII, showed the transgene in GUS negative plants to be highly methylated relative to the transgene in GUS positive plants. Some of the studies on gene silencing also indicate that multiple integration pattern and copy number is also associated with DNA methylation.

Cannell and coworkers studied the inheritance of gus and bar marker genes over three generations. The integration, inheritance and expression of these marker genes in the population studied were stable and predictable with a few exceptions. The inheritance of integration patterns was stable, and transmission/ inheritance of the transgenes followed Mendelian ratios in a majority of lines. From this study, the authors claim that the transformation procedure, transgene integration, and marker gene expression had little effect on the transmission of transgenes to the subsequent progenies. However, over the three generations studied, a uniform, 'progressive' transgene silencing, specific to the gus gene, was observed. This gradual loss of the gus gene expression was not accompanied by a reduction in bar expression. Since in this case, the gus and the bar genes are located on the same plasmid, observations indicate a post-transcriptional gene silencing.

The silencing of HMW glutenins was also observed in transgenic wheat expressing extra HMW subunits. Characterization of six independent events involving the transformation of wheat HMW-GS genes, 1Ax1 and 1Dx5, in a cultivar expressing five subunits by particle bombardment resulted in partial or complete gene silencing. Silencing of all the HMW glutenin subunits was observed in two different events of transgenic wheat expressing the 1Ax1 subunit transgene and overexpressing the 1Dx5 subunit. Control of gene silencing thus remains a challenge for the immediate future. Meyer, (1995) advocated that the problem of gene silencing in wheat can be minimized by optimising methods for simple integration patterns, use of promoters and gene sequences isolated from cereals, use of matrix associated regions (MARs) or scaffold attachment regions (SARs), which insulate transgenes from surrounding chromatin, might help in reducing the gene silencing problem.

TRANSPOSON TAGGING IN WHEAT

Transposon mutagenesis has been widely exploited in various organisms to isolate genes that encode unidentified products. The maize activator (Ac) and Dissociation (Ds) elements are the best-studied transposable elements in heterologous host plants. Initial studies have reported the introduction and activation of the Ac/Ds elements into cultured wheat cells by using a wheat dwarf virus by particle bombardment. Takumi and coworkers has developed a transposon tagging system in wheat by introducing the Ac transposase gene under the CaMV 35S promoter into cultured wheat embryos by particle bombardment.

For the development of a transposon tagging system, embryos isolated from a stable Ac line were bombarded with a plasmid containing the maize dissociator (Ds) element located between the rice Act1 and the gus genes. The transient expression of the gus gene was observed after the excision of Ds elements. Southern and northern analysis of the T0 and T1 plants has exhibited the stable expression and inheritance of the Ac transposase gene. These results have demonstrated the precise processing of the maize Ac transposase gene and the synthesis of the transposase protein in transgenic Ac lines. The Ac transposase gene causes the transactivation and excision of the Ds in the Ac transgenic lines transformed with the maize Ds elements. Thus in near future we can expect traits of commercial importance to be tagged for a wider utility in important crop plants.

MARKER ASSISTED SELECTION IN WHEAT BREEDING

Conventionally, plant breeding depends upon morphological/phenotypic markers for the identification of agronomic traits. With the development of methodologies for the analysis of plant gene structure and function, molecular markers have been utilized for identification of traits. Molecular markers act as DNA signposts to locate the gene(s) for a trait of interest on a plant chromosome, and are widely used to study the organization of plant genomes and for the construction of genetic linkage maps. Molecular markers are independent from environmental variables and can be scored at any stage in the life cycle of a plant. Over the last several years, there has thus been marked increase in the application of molecular markers in the breeding programmes of various crop plants. Molecular markers not only facilitate the development of new varieties by reducing the time required for the detection of specific traits in progeny plants, but also fasten the identification of resistance genes and their corresponding molecular markers, thus accelerating efficient breeding of resistance traits into wheat cultivars by marker assisted selection (MAS). The availability of back-cross derived near isogenic lines (NILs) have also facilitated the analysis of various lines by using different marker systems. The introduction of alien genetic variation into wheat is a valuable and proven technique for wheat improvement. Wild relatives of wheat provide an enormous resource of new genes for wheat improvement, particularly disease and stress tolerance. These genes can be of use if recombined into lines adapted to the conditions of a particular region.

Initial studies on the application of molecular markers in wheat relied on the hybridisation based restriction fragment length polymorphism (RFLP) system-RFLP maps provided a more direct method for selecting desirable genes via their linkage to easily detectable markers thereby expediting the movement of desirable genes among varieties. The factors that had been instrumental for the use of RFLP in wheat was the limited number of polymorphisms observed among wheat lines and more significantly the availability of aneuploid stocks for the determination of chromosomal location of genes. In wheat, RFLP's have been used to map seed storage protein loci, loci associated with flour colour, cultivar identification, vernalization (Vrn1) and frost resistance gene on chromosome 5A, intrachromosomal mapping of genes for dwarfing (Rht12) and vernalization (Vrn1), resistance to preharvest sprouting, quantitative trait loci (QTL's) controlling tissue culture response (tcr), nematode resistance, milling yield, resistance to chlorosis induction by Pyrenophora tritici-repentis.

RFLP markers are also useful in selection programmes for resistance against pests and pathogens, which is otherwise labour and time consuming, and to detect homozygous individuals and have been used for resistance to barley yellow dwarf virus, resistance to wheat spindle streak mosaic virus, resistance against powdery mildew, resistance against leaf rust, resistance against cereal cyst nematode. The use of RFLP analysis in wheat has, however, been of limited use in the intervarietal analysis due to low level of polymorphism and the high cost for screening in breeding situations.

With the development of polymerase chain reaction (PCR) methodologies, Random Amplified Polymorphic DNA (RAPD) emerged as a convenient and effective technique for tracing alien chromosome segments in translocation lines. RAPD markers provide a useful alternative to RFLP analysis for screening markers linked to a single trait within near isogenic lines and bulked segregants. He *et al.* (1992) reported the development of a DNA polymorphism detection method by combining RAPD with DGGE (denaturing gradient gel elecrophoresis) for pedegree analysis and fingerprinting of wheat cultivars. RAPD markers can be converted to more user-friendly Sequence Characterized Amplified Region (SCAR) markers, that display a less complex banding pattern.

SCAR markers linked to resistance genes against fungal pathogens have been characterized in combination with RAPD and RFLP. In recent years, RAPD and other PCR based markers like Sequence Characterized Amplified

Regions (SCAR), Sequence Tagged Sites (STS) and Differential Display Reverse Transcriptase PCR (DDRT-PCR) are increasingly being used for identification of desirable traits in wheat and related genera. These markers have been used in particular for disease resistance against viral and fungal pathogens and also for insect and nematode pests and have the potential of pyramiding of resistance genes for effective breeding programmes.

PCR based markers have been extensively characterized for genes of resistance against common bunt, Tilletia tritici, powdery mildew, Erysiphe graminis, leaf rust, Puccinia recondita, resistance against Hessian fly, Mayetiola destructor and Russian wheat aphid, Diuraphis noxia.

Simple sequence repeats or microsatellites are more promising molecular markers for the identification and differentiation of genotypes within a species. The high level of polymorphism and easy handling has made microsatellites extremely useful for different applications in wheat breeding. Microsatellites have also been used to identify resistance genes like Pm6 from Triticum timopheevii and Yr15 from breadwheat. In near future, molecular markers can provide simultaneous and sequential selection of agronomically important genes in wheat breeding programmes allowing screening for several agronomically important traits at early stages and effectively replace time consuming bioassays in early generation screens.

2

Benefits of Urban Forests and Trees

Trees in the urban environment contribute significantly to the aesthetic appeal of cities, helping to maintain the psychological health of the inhabitants. Besides the aesthetic and environmental aspects, urban forestry is also of importance in helping populations poor in resources to meet basic needs. Urban plantings of fruit trees to provide firewood is one way of utilizing obvious benefits from the available natural resources, as done in Kampala, the capital of Uganda. Research has shown that urban trees benefit communities economically, socially and environmentally. The size, structure and condition of the urban forest directly affects the amount of benefits provided by the urban vegetation. Nevertheless, the social and environmental benefits provided by the urban forest are often largely ignored in land-use planning.

ECONOMIC IMPACT

Though trees are not usually thought of as economic resources, the presence of trees has an impact on different economic factors, including the level of real-estate prices, economic investments and employment. Results of recent Finnish studies show that the benefits provided by urban forests are reflected in property prices, and that environmental variables, such as proximity to wooded recreation areas and water courses as well as an increasing proportion of total forested area in the housing district, had a positive influence on apartment prices. The presence of trees may influence property values positively by as much as 20%, with an average increase of 5-10%. Trees also provide different external environmental benefits. Costs for heating and cooling can be reduced by appropriate use of vegetation. Energy reductions for individual buildings have been shown to range from 5 to 15% for heating and 10-50% for cooling. Especially in areas with high summer temperatures, trees are important providers of shade. In areas with cool winters, it is of importance to use deciduous trees, as the loss of leaves in winter allows the sun to heat the house. Proper arrangements of vegetation around buildings can also reduce wind velocity and thereby reduce the heat loss from buildings.

RECREATIONAL USE OF GREEN AREAS

Several studies have shown the importance of urban forests for participation in outdoor recreation activities. In a Danish national survey it was found that around two-thirds of all forest visits take place in the forest situated nearest the housing area. Danish and Swedish studies have also showed that the use of green areas by people is associated with three highly esteemed values: (i) a high degree of natural elements, (ii) fresh air and (iii) the possibility of solitude. In general, people use green areas less than they would like. According to Swedish studies, distance and the fear of assault are the most prevalent reasons for people not visiting green areas. When the distance to the park exceeds 300 m, one person in four postpones a daily visit. As many as 56% refrain from regular walks in the park when the distance increases to 500m. American studies indicate that the removal of vegetation at strategic points can reduce the risk of assault, although this should not be the reason for removing all the vegetation.

Psychological Aspects

City life is stressful but research shows that urban green areas have a beneficial influence on the health and well-being of the urban population. Studies indicate that visits to green areas can counteract stress, renew vital energy and speed healing processes. In Sweden, Grahn (1989) conducted extensive studies on the significance of parks to different groups of the population. For instance, he persuaded 40 schools, hospitals, sports associations, cultural associations and day-care centres to keep diaries of their outdoor activities for 1 year. For example, the diaries show that periods spent outside had an actual medicinal value for patients and residents of hospitals, old people's homes and homes for the sick. People became happier, slept better, needed less medication, were less restless and far more talkative.

Ulrich (1984) showed that hospitalized patients recovered faster when they had a view through a window, allowing them to see trees. Ulrich *et al.* (1991) showed a gory film on industrial accidents to 120 people. Half the people were then shown a nature film, whereas the other half were shown a film on the city, with sequences of buildings and traffic. The subjects' heart beat, muscular tension and blood pressure were monitored throughout. All subjects exhibited strong signs of stress during the first film on industrial accidents. The stress levels of the subjects that then watched the nature film returned to a normal level after 4-6 min, whereas the half that watched the film on buildings and traffic continued to exhibit high stress levels. Kaplan and Kaplan (1989) have formulated a theory on the interaction between human attention and the surroundings. This theory distinguishes between spontaneous attention and conscious attention. Spontaneous attention demands no effort and occurs without premeditation. Conscious attention demands energy and leads to psychological exhaustion in the long term. Rapid movements, strong colours, sudden noises and strong odours are typical stimuli that demand conscious attention. These signals have a powerful effect on our attention, consciously and unconsciously, because they are perceived as potential dangers to which we should react. This means that urban living, with fast vehicles, flashing neon signs and strong colours, causes constant stress. Kaplan and Kaplan's research indicates that vegetation and nature reinforce our spontaneous attention, allow our sensory apparatus to relax and infuse us with fresh energy. Visits to green areas bring relaxation and sharpen our concentration, since we only need to use our spontaneous attention. At the same time, we get fresh air and sunlight, which have significance for our diurnal and annual rhythms.

Environmental Education

The change and the continuity in nature provides not only a sense of time but also a sense of security and confidence, through the predictable change of seasons and the repetition of natural processes. The urban trees and forests play a role in educating young people and children in understanding the basic processes in nature and the complexities of the environment, either informally during play or as part of the school curriculum. Playing in nature or natural settings also helps children to develop their motoric senses and to regain and experience the human connection to nature through their imagination. Experiences with environmental education in the USA show that access to even a small area of the urban forest is of importance when learning about natural processes. In England in 1985 a research project supported by the Department of Education and Science, the Countryside Commission and a consortium of local authorities resulted in the organization Learning through Landscapes (LTL). LTL was founded on the recognition of the impact that the physical surroundings has on children. LTL helps schools to design schoolyards that improve the quality of the environment and helps to create additional resources for the formal curriculum, such as the promotion of more effective teaching and learning in the outdoor classroom.

The English Community Forests are examples of integrated programmes in which environmental education is integrated in social programmes by well-coordinated planning. The forests are charged with running and coordinating social programmes, involving information, consultation, participation, art, education, sport, recreation and cultural activities.

Community Involvement

Other benefits of urban forestry are those provided by involving people in planning, planting and caring for the trees and urban forests in their own locality. By promoting social interaction and strengthening local pride and identity, local public engagement benefits management greatly, encouraging people to act protectively towards their local forests and providing a potentially huge volunteer workforce. The possibilities for creating public awareness in any town are endless. In the USA there is a tradition of involving the public in all processes, from the policy-making to the care and maintenance of the urban forest. Through various types of tree programmes, public awareness has proved to be a valuable tool in the justification of tree activities because it develops a public interest group that helps to lobby for support. The projects that involve citizens have various aspects in common, such as planting, preserving, pruning, maintenance, public education, etc.. Examples of initiatives that enhance public awareness and engagement in a municipal tree programme are the Tree Boards, municipal non-profit groups with participants from a wide range of professions. The close cooperation between professionals and non-professionals provides direct communication of new ideas and research and helps to solve associated problems as experienced in a practical situation. Another way of involving the public is to encourage houseowners to adopt a tree, by taking part in the care and maintenance of a tree in their neighbourhood.

URBAN GROWING CONDITIONS

Urban growing conditions differ significantly from those in the rural landscape, and produce difficulties as a result of both above- and below-ground influences.

STRESS FACTORS

The harsh soil and air conditions that exist in urban planting are problems that do not play the same role in landscape planting. Growing conditions may also be difficult due to shading effects, recreational users, etc.. The modified urban mesoclimate affects the quantity of contaminants in urban areas, which is raised by a factor of around 25. In general, the average lifespan of a newly planted street tree may be as low as 10-15 years. During the last 30-40 years, the vitality of street trees has fallen drastically. Heavier traffic patterns have increased demands for road construction, which consequently has changed the growing conditions of many roadside trees. Also, pollution from traffic has a highly detrimental impact on street trees. The fact that 50% of the trees planted in an urban environment die within the first year emphasizes this point. Nowak *et al.* (1990) found that 34% of 480 trees died within 2 years of planting, while Miller and Miller (1991) found that the mortality rate was 25-50% for a number of species planted in Wisconsin, USA.

Temperature extremes can occur, especially where trees are widely spaced and where heat is reflected from hard surfaces. Harris (1992) described that, occasionally, tree limbs up to 0.6 m and trunks up to 1.2 m in diameter break and fall during hot calm summer and autumn afternoons and subsequent evenings. Roots are more sensitive to temperature extremes than the tops of plants. Wind speed will vary according to the shape and height of buildings. Areas with tall buildings will usually be relatively cool in summer due to shading effects, and warmer in winter due to wind-protection effects. On the other hand, winds are more variable and more extreme at exposed corners of tall isolated buildings. Buildings deflect strong winds downwards and concentrate their force at the base and corners of buildings, forming 'wind tunnels'. Trees planted in these exposed gaps may suffer scorched leaves and shoots, which lead to a stunted canopy, especially on the windward side. Newly planted trees will transpire more rapidly in windy situations, which can lead to the death of a tree already severely stressed by drought. The wind stability of trees is determined by tree species, stand structure, spacing, thinning regimes, soil classes, breeding and tree age at the time of anchorage.

The presence of airborne pollutants in the atmosphere has been a characteristic feature of the urban environment since the beginning of the Industrial Revolution. Air pollution can occur in a variety of forms but the principal ones are dust, SO_2 and NO_x. Leaves are the plant parts most likely to show symptoms of air pollution injury. On broadleaved plants, leaves may develop interveinal necrotic areas, marginal or tip necrosis, stippling of the upper surface, or silvering of the lower surface.

However, trees in the urban environment also play a role in the quest for cleaner air in the cities. Scott *et al.* (1998) showed that daily uptake of NO_2 and particulate matter represented 1-2% of anthropogenic emissions for the county of Sacramento, California. In areas with winter temperatures below 0°C, the use of de-icing salt is a well-known problem. De-icing salt is applied to the surrounding environment by surface run-off, wet spraying and airborne drifting. The initial and most common symptom of de-icing salt damage on trees and shrubs is reduced growth. This is often difficult to recognize or may be confused with other stress factors. Reduced growth is usually followed by early autumn colours and premature leaf fall. De-icing salt is usually accumulated on the windward side of trees. The damage is easily recognized because it faces the road and is normally regarded as the best indication of de-icing spray damage. The majority of trees and shrubs subjected to either soil salt or salt spray typically show necroses at the edges of the leaves or needles. Wounds, often related to pruning, are a common place for spray salt to infect the plants. The damage may cause lack of sprouting and eventually dieback. Conifers are very susceptible to de-icing salt spray damage because they are green all through the winter maintenance season. Trees and shrubs damaged by de-icing salt and showing dieback are difficult to cure.

Characteristics and Restriction of Rooting in the Built Environment

Urban soils as a growing medium are poorly understood and often misunderstood. Therefore plantings are carried out with little appreciation or attention to the character and quality of the material that lies beneath the surface. One major problem in relation to planting in the urban situation is soil compaction, which may occur in small as well as large urban sites. Soil compaction can be divided into two types: (i) intentional soil compaction, which occurs when soil is deliberately compacted for site stabilization under roads, houses, etc. and (ii) unintentional soil compaction, which occurs when traffic uses areas intended for planting (Randrup 1997). In the urban situation, unintentional soil compaction is primarily found along roadsides and on construction sites.

When soil is compacted, its bulk density increases and its porosity decreases. These effects inhibit plant growth because the soil becomes impenetrable to root growth and, furthermore, restricts the water and oxygen available to the roots. For example, root growth of most plants is impeded once soil bulk density rises above 1.6. One consequence of compacted soil is waterlogging, which can kill roots around existing trees. Soil loosening has proved to be effective in alleviating compacted soil. However, there is no doubt that the best treatment for compacted soil is to protect the soil from being compacted in the first place. Florgård (1987) suggested protecting trees growing on construction sites by dividing the site into zones in which different types of construction traffic are permitted. The principle of construction site zoning was adapted by Randrup and Dralle (1997) to protect the soil from being compacted. They suggested that the entire construction site be divided into a building zone, a working zone and a protection zone.

URBAN CHARACTERISTICS/ENVIRONMENT ANALYSED

The city is characterized by paving and buildings, which variously results in decreased wind speed, but an increase in windthrow, raised temperatures and precipitation, lowered humidity and shading in street canyons. The extent of the influences of these factors depends on the structure and amount of vegetation, as well as on the size of the city. Green areas in the city accessible to the public include areas at schools, public libraries and social institutions, as well as parks and churchyards. These are areas that are typically relatively small and geographically widespread, which is why they often suffer great recreational pressure, most often due to the overall proportion of green areas in the city.

Throughout Europe the proportion of urban green areas varies greatly, from over 60% of the area of Bratislava, the capital of Slovakia, to about 5% in Madrid, the capital of Spain. In comparison, the figure for Mexico City is only 2.2%; in relation to the number of inhabitants, this only provides 1.94 m^2 per inhabitant, which is far below the 9 m^2 recommended by the World Health Organization. A suggested measure of urban environmental quality is the location of green areas within a walking distance of 15 min or less from all housing areas. This criterion is met for all citizens in Brussels, Copenhagen, Glasgow, Gothenburg, Madrid, Milan and Paris and, in general, for more than 50% of the population in most European cities. The quality of urban green areas are increasingly recognized as being important to the overall quality of human life in the cities. Beyond this, the urban forests and trees are important as ecosystems in relation to the conservation of biological diversity. Though the urban population benefits from the urban green areas, the increase in population places great pressure on the existing green areas as a result of urban and infrastructural development. The importance of the urban green space, of which urban forestry is an integral part, increases as population increases. The growing urban population needs the environmental and social benefits associated with urban forests, as shown by Ulrich (1984) and Grahn (1989). In developing countries, urbanization has had a dramatic influence on creating environments practically without any amenities, as is the case in Mexico City, where the growth in population has not been matched by an increase in green space. It is an accepted reality that the growth of the cities cannot be stopped. Instead, the challenge is to control urban growth so that it results in economic growth and a satisfactory environment.

As more than two-thirds of the population of Europe live in urban areas, the quality of the urban environment, including green areas, is becoming increasingly recognized as one key to the economic reconstruction of European cities (Commission of the European Community 1990). Urban areas constitute the everyday environment of the greater part of the population and, in recent years, this topic has received considerable attention in the European Union, United Nations and the Organization for Economic Cooperation and Development, which will hopefully lead to an improvement in our knowledge of urban environments.

TREES IN THE URBAN ENVIRONMENT

Trees in the urban environment are often referred to as the urban forest, comprising trees in civic woodlands, parks and the street. Earlier, urban trees were mainly regarded as aesthetic elements, whereas today they are recognized as having a positive impact on the environment as well as providing economic and social benefits. Monetary evaluations reflect the various benefits arising from the urban forest, covering such aspects as reduction of pollution and energy use, environmental amelioration, savings in public health care and increase in economic investment. Hence, the value of the urban forest is being increasingly recognized as a vital component in the maintenance of a sustainable urban environment in cities around the world. Meanwhile, the population living in urban areas has increased rapidly since the 1950s and the lack of space makes it tempting to use green areas for infrastructure and buildings.

Collins (1997) emphasizes that urban forestry, the planning, management and maintenance of the urban forest, is more closely aligned to traditional forestry than might be immediately apparent. As implicit in the term, the principle of sustained yield has been adapted to the urban environment, applying rural land-use forestry to the management of the urban forest. Hence, the overall objective in urban forestry is not that of timber production but, through a balanced structure of age and species, a sustained production of environmental, social and economic benefits. These social benefits also accrue if local communities are encouraged to contribute to their own environment by promoting projects and activities involving local residents. These kinds of projects often prove effective in promoting social interaction and lead to increased local involvement in other aspects as well.

The planning and management of a healthy urban forest requires that coordinated strategies are agreed between the professions more or less directly involved, such as planners, landscape architects, arboriculturists, foresters, engineers, legislators, developers and utility managers. Management of the urban forest resource is moving towards the achievement of a healthy, well-distributed urban forest, which is increasingly becoming an essential and integral part of the urban infrastructure.

Trees in the urban environment have been defined in several ways, although 'urban forestry and arboriculture' is probably the most used term in relation to trees in or near the urban environment. Many different research disciplines are involved in the fields of arboriculture and urban forestry. Harris (1992) defines arboriculture as being 'primarily concerned with the planting and care of trees and more peripherally concerned with shrubs and woody vines and groundcover plants'. However, many people consider urban forestry to consist primarily of two types of planting: urban forests and urban trees. Also, definitions of urban green areas and urban forestry vary significantly throughout the world. There seem to be different opinions as to what urban forestry covers, depending on whether professionals have a background within or outside forestry or have experience of working in the USA or other parts of the world such as Europe.

In many cases foresters have argued that urban forestry concerns 'forestry in urban areas'. Volk (1986) stated that:

Green areas such as tree lined streets, cemeteries, and parks, which are not predominated by trees and as such do not come under forestry influence should be excluded from consideration [of being considered urban forestry]. On the other hand, park forests which come under forestry supervision, should be included in our considerations. The fear of involvement of foresters in urban plantings has been described by Chambers (1987): 'Misguided resistance to the concept [of urban forestry] is still often mistakenly assumed to imply the take over of public open space and existing amenity trees for timber production.' Americans tends to look at urban forestry as the 'management of trees in urban areas on larger than an individual basis'. In Europe this broad concept now seems to be accepted on a wider basis. Urban forest stands are now often considered as resources where an economic yield is not required, and traditional forestry practices are combined with both aesthetic and recreational considerations.

Costello (1993) suggested that urban forestry be defined as 'the management of trees in urban areas', including single trees. In this definition, 'management' is described as the planning, planting and care of trees; 'trees' as individuals, small groups, larger stands (*e.g.* green belts) and remnant forests; and 'urban areas' as those areas where people live and work. The location of urban forests can be anywhere, from urban settings to the countryside, as long as there are human structures on the site; these structures can be related to ecological functions, protection, merchandise or tourism and recreation. Using this concept, we categorize trees in the urban environment as both urban forests and urban trees. Urban forests can be defined by their placement in or near urban areas and by their multifunctional aspects, giving shade, amenity values, etc. Therefore, we define urban forestry as the establishment, management, planning and design of trees and forest stands with amenity values, situated in or near urban areas.

Trees in the urban environment may be divided into three different types: trees in urban woodlands, park trees and street trees. These types of plantings differ distinctively in several ways. Most importantly, they have significantly different growing conditions, which means that they have different needs with regard to planning, management and maintenance. The methods used and the research problems related to urban green areas are common throughout the world due to the fact that urban 'greening' concerns not only natural spheres (such as trees, growing conditions, etc.) but also the environment in close proximity to human populations.

3

Forest Tree Improvement and Breeding

Breeding involves crossing among selected trees to produce progenies with new combinations of genes for testing and selection. The resulting selections will be the clones of the next generation of improvement. Breeding requires flowering trees of the selected clones that will be used as parents. This is a major problem for exotics like P. deltoides, since almost no provisions have been made to establish and maintain breeding orchards of mature individuals of the introduced clones.

All controlled crosses involving P. deltoides before 1997 have been opportunistic, depending on what was flowering. Seven P. deltoides clones flowered during this time (G-48, D-121, and S7C8 were females, and G-3, S7C1, S7C15, and S7C20 were males). Except for D-121, all of these were from Brazos County, Texas. One male flowering clone of P. yunnanensis and one male of P. `robusta' (Euramerican hybrid) were also used. "Breeding" in U.P. has been by the U.P. Forestry Department's Lalkuan Research Centre, Silviculture Division, Sal Region, by nearby WIMCO Seedlings Ltd. in Rudrapur, and at FRI in the Genetics and Tree Propagation Division. Initially, most of the new clones came from open pollinations between G-48 and G-3 (the first two clones to flower).

These seedling progeny clones provided the "L-series" clones of the U.P. Forestry Department and most of the "WSL-series" clones of WIMCO. Subsequently, the U.P. Forestry Department nursery manager at Lalkuan made the following crosses (female x male): G-48 x G-3, D-121 x S7C1, G-48 x P. ciliata, D-121 x P. ciliata, G-48 x P. yunnanensis, and G-48 x P. `robusta'. WIMCO selected three seedling clones from their own controlled crosses of G-48 x G-3, and they made some backcrosses (such as G-48 x P. `robusta) to stabilize desired hybrid combinations of P. deltoides with P. nigra. Controlled crosses by the Noh, E.R., K.K.Sharma, and M.L.Kapoor. 1995. A report submitted to FAO on Improvement Programme of Indigenous Poplars with Particular Reference to India. FRI, Genetics & Tree Propagation Division, Dehradun. Genetics and Tree Propagation Division at FRI were made in the spring of 1996 and required transport of flowering branches from Lalkuan.

The crosses G-48 x G-3 and G-48 x P. ciliata were successfully accomplished. The P. ciliata male flowers came from the hills of nearby Mussoorie.

The greatest amount of controlled crosses among Populus species in India has taken place at the Y. S. Parmar University of Horticulture and Forestry in Solan, H.P. Flowers (male and female) of P. ciliata could only be successfully ripened on cuttings in the greenhouse if the cuttings were cleft-grafted to potted rootstock seedlings in January before bud break. Water cultures and 'twig in pot' methods did not work. Cuttings of P. gamblei bearing male flowers could not be forced to ripen and shed pollen at Solan under any method, so no crosses with this species were made. Dr. Khurana successfully produced the hybrids P. ciliata x P. maximowiczii (pollen obtained from Japan), P. ciliata x P. yunnanensis, and P. deltoides (female) x P. ciliata (male) (but not the reciprocal) (personal communication).

The P. ciliata x P. maximowiczii hybrid was heterotic for growth, but had the branch knot problem of P. maximowiczii. Dr. Khurana selected 20 clones from field tests of this hybrid for minimum branch knots.The P. ciliata x P. yunnanensis and P. deltoides x P. ciliata hybrids did not show heterosis for growth in his tests. Backcrosses of the Euramerican hybrid clone '1-455' (female), which exhibits fast growth but is highly susceptible to Melampsora spp. of leaf rust, with P. deltoides (male) by Dr. Khurana produced two selections that exhibited extremely fast growth (4 to 5 meters per year in height) on recently-exposed, well-drained forest sites (not agroforestry) in H.P.

A total of 358 clones (primarily P. deltoides, including some intraspecific crosses of G-48 by G-3) have been preserved in a germplasm bank at FRI in Dehradun, U.P. There are also 277 clones of P. deltoides in a germplasm bank at the HFRI Shilly Research Nursery in Solan, H.P. Most of these are duplications of the clones at FRI, so that protection against a catastrophic loss of clones (from fire, etc.) is provided. The U.P. Forest Department's Lalkuan Research Centre is also maintaining a collection of clones from previous introductions and from inter-and intraspecific crosses. Many of these are duplications of clones at FRI. There are 150 clones in the Department's nursery at Lalkuan (100 P. deltoides clones), another 60 clones in a replicated test planted in February 1985 at `Ganga Pur Patia East' near Lalkuan, and 233 clones in the Department's populetum planted in February 1989 at Tanda, U.P. (Plot #47). Forty-four of the clones in the Tanda planting came from Dr. Hansen's 1986 shipment of 200 clones to India and were selected based on nursery performance at the Lalkuan nursery. The Ganga Pur and Tanda tests were just beginning to flower in 1996 and will provide the only breeding orchard available in India (108 clones) for the initiation of an advanced-generation mating design with P. deltoides.

NURSERY AND GREENHOUSE TECHNIQUES

All Populus propagules being commercially planted in agroforestry or in reforestation/ afforestation are E.T.P.s (Entire Transplants). These are one-year-old rooted cuttings. The tree improvement programme must therefore use E.T.P.s for clonal field trials. Nursery management techniques include soaking the cuttings and nursery beds in water before planting and then planting the 18-25cm length cuttings at a spacing of 80cm x 60cm. Debudding to remove young limbs on the E.T.P.s is done from June to November. The nursery beds are flood irrigated 1-2 times per week during March-June, until the rainy season begins. Backpack sprayers are used to apply insecticides for control of leaf beetle, thrips, and wood-borer insects.

The target E.T.P. for production nurseries is a plant whose ground-line diameter is 1/100th of the height (in cm) (personal communication with Dr. J. P. Chandra). When harvesting the one-year-old E.T.P.s in December-February, the tap root is cut at a 25cm depth and all side roots more than 10cm long are trimmed. The plants are packed, transported, sold, and planted with a naked root. Research nurseries have the additional task of developing uniformity among E.T.P.s of the same clone. Re-propagating E.T.P.s annually from the previous year's E.T.P.s for 4-5 years in the nursery may be required to remove "C-effects" (age, position-in-tree, or vigor effects on cuttings taken from the crowns of different-aged ortets on different sites). The clones will not be taken to the field trials until within-clone variation is less than 10-15% (personal communication with Dr. Khurana at Solan). Furthermore, the multiplication of seedling-derived clones will take at least three years in the nursery to produce enough E.T.P.s for field trials at 2-3 sites. When seeds are produced from controlled crosses during breeding, they must be germinated and grown before they can be vegetatively multiplied as E.T.P.s. P. ciliata seeds are very fragile and will lose viability in 7-10 days if not germinated quickly (personal communication with Mr. D. V. Negi).

All Populus species seedlings are vulnerable to sun scald (heat) damage and diseases (damping off) during germination. Procedures for germinating and growing seedlings at WIMCO Seedlings Ltd. provide a working model for success. Fungicide-treated seeds are planted in rows on moist, heat-sterilized sand in flat clay pots and placed in a double-walled polygreenhouse with fans. The polygreenhouse is covered with 50% shade cloth. Approximately 15 days after the seeds have germinated, the germinants are 'pricked' and transplanted to individual cells in container racks. The soil in the containers is a barnyard mixture of manure that has been sterilized, so no additional fertilization is required. Containerized transplants are placed in a mist chamber in the greenhouse for one week (mist for 1-2 seconds at 5-minute intervals), then moved into a shadehouse for one week and subsequently outside under the shade of trees for 15 days.

Finally, the acclimated seedlings are placed in the open sunlight where they remain until they are 30cm tall. At this size the roots can retain soil in a 'plug' when removed from the container. The 'plug' seedlings are then planted in the research nursery (personal communication with Dr. S. Chauhan of WIMCO).

AGROFORESTRY

The entire agroforestry industry with Populus is founded on a very limited genetic base. Approximately 90% of poplars being planted for agroforestry in U.P., Haryana, and Punjab come from the P. deltoides clones G-48, G-3, and S7C15 (personal communication with Dr. J. P. Chandra of WIMCO Seedlings Ltd.). The next two clones in popularity are 'Udai' from WIMCO and L-34 from the U.P. Forestry Department. Both of these clones come from open pollinations between G-48 and G-3. Other clones that are not quite as desirable, but which are being held in reserve in case of a serious disease or insect outbreak, are S7C8, S7C4, L-43, ST-240, ST-70, and ST-67. All of these clones (except the three ST clones) came from gene pools in Brazos County, Texas. The three ST clones came from the southern Mississippi River alluvial plain just north of Vicksburg, Mississippi USA.

Typical agroforestry methods are to plant 4m-5m tall E.T.P.s in (1) "bund" (shelterbelt) plantings on the small levees around flood-irrigated agricultural fields, (2) "block" plantings at 5m x 5m, 5m x 4m, or 6m x 4m spacings within agricultural fields (underplanted with the crops), or (3) "row" plantings that contain alternating rows of Poplars and horticultural tree species underplanted with agricultural crops. Typical spacing between "rows" is 6m, so the P. deltoides rows are 12m apart. Rotation lengths for poplars are 5-8 years (5 years for plywood, 8 years preferred for veneer and matches). Typical crop combinations with "bund" plantings are sugarcane or rice, but rice is not recommended because of the requirement for summer flooding.

Crop combinations with "block" plantings involve sugarcane for the first two years and then winter wheat or a combination of winter wheat alternating with summer-grown tumeric or pearl millet during the third through eighth years of the P. deltoides rotation. A new alternative to tumeric or pearl millet is celery. Dr. N. B. Singh has also observed rice being cultivated in some block plantings during the fourth through eighth years. Apparently, P. deltoides can tolerate flooding at the older ages. "Row" planting mixtures in agro-horti-forestry systems are based on a rotation length of up to 50 years for the final horticultural orchard. These systems may involve poplar-peach, poplar-litchi, or poplar-mango combinations with sugarcane, tumeric, and winter wheat as under crops. Up to two 5-to 8-year rotations of Populus may be grown before the fruit trees (particularly mango) spread in crown diameter to become a full orchard. In all of these systems the poplar trees are pruned of lower limbs (recommended lower 1/3 of stem height after the second year). The leaves are collected and composted at the end of the growing season for placement back on the fields, and the stumps are dug up and removed for fuelwood/charcoal at the end of the rotation.

UNDERSTANDING THE TREE IMPROVEMENT

Tree improvement relies on understanding and using variation that naturally occurs in tree populations. Tree improvement increases the value of a tree species by 1) *selecting* the most desirable trees from natural stands or plantations, 2) *breeding* or mating these select trees and 3) *testing* the resulting progeny. The trees involved is this process are referred to as the*breeding population.* This three-step process is then continuously repeated to further improve the average value of the breeding population. Each iteration of this improvement process is referred to as a *generation.* The *production population* is another group of trees established to meet commercial planting demands. Usually the production population is a *seed orchard* or group of trees at a single location managed specifically for seed production. For species which can be readily propagated using rooted cuttings, a *hedge orchard* managed to produce cutting material may serve as the production population. Seed and hedge orchards generated from a tree improvement programme are established from grafts, rooted cuttings or seed of the very best trees in the breeding population.

SELECTION

Tremendous variation exists in natural stands for many important Christmas tree characteristics such as growth rate, colour, branching habit and disease resistance. Growers often encounter this variation in their Christmas tree plantations. Observed characteristics such as growth rate are referred to as the *phenotype* of a tree. A tree's phenotype is determined by both its genetic constitution, or *genotype* and the effect of the *environment* as described by the following equation:

$$P = G + E$$

where,

P = phenotype
G = genotype
E = environment.

When selecting trees, what you see is not necessarily what you get. Genetically inferior trees may sometimes appear phenotypically desirable because they grew in an unusually favorable micro-environment. Conversely, genetically superior trees may appear phenotypically undesirable due to poor environmental conditions. Characteristics vary in the degree of genetic versus environmental influence. For example, genetic control of branching habit is stronger than for height growth so that the selection process is generally more effective for branching traits. Initial selection of trees from natural stands or unimproved plantations are necessarily based solely on phenotype. Since the variation due to environmental effects is less in plantations than natural stands, selection is more effective in plantations.

In order to ensure that the selection process is as effective as possible in unimproved stands, candidate trees are graded according to predetermined criteria. Both the candidate tree and neighboring check trees are measured and compared in order to account for local environmental effects. Once a tree improvement programme is underway, genetic information (performance of relatives) is used to increase the effectiveness of the selection process.

BREEDING

The next step in the tree improvement process is mating or breeding among the select trees. For this purpose, branch tips or *scion* of each selection are grafted onto seedling *rootstock* to establish a *breeding orchard. Control-pollinations* are then performed among the selections. Control-pollination starts with covering receptive female cones with bags during early spring to prevent natural pollination. Pollen collected from other select trees or mixed from a group of select trees is then injected into the pollination bag. After pollination has occurred, the bags are removed and the cones allowed to ripen. Each cone-producing branch is labeled to ensure correct identification at cone harvest. Control-pollination is performed according to certain*mating designs* among the select trees in order to maximize the overall genetic information derived from the resulting progeny.

TESTING

Seed produced from tree improvement breeding efforts are used to establish progeny tests. The purpose of these tests are to 1) provide genetic information about the select parent trees and 2) provide an improved population of trees from which the next generation of select trees is made. Tree families established in progeny tests are randomized and replicated in a designed manner to meet statistical and genetic criteria. Families are planted in several sites and in more than one year to sample an adequate number of environmental conditions. Christmas tree progeny tests are intensively managed similarly to a typical Christmas tree plantation. Growth and quality measurements are made periodically in these tests until harvest. These data are entered into a data base and analyzed. The results are used to assess the genetic worth of the original selections, to make selections for the next generation and to make recommendations for establishing and upgrading seed and hedge orchards.

Final Remarks

Tree improvement is the successive application of selection, breeding and testing to continuously improve the value of a population of trees. The best trees of the population are used to propagate planting stock for commercial plantations. While the Christmas tree industry stands to reap tremendous benefits from tree improvement, it should not be viewed as a panacea. Tree improvement is one of many tools needed to continue to improve the productivity and quality of Christmas trees crops. Growers must continue to use the best available technology to manage their plantations. In fact, the greatest benefits from tree improvement will be achieved on the best sites receiving the best management.

Due to the nature of trees, the tree improvement process requires both a large-scale effort and a relatively long time-frame to achieve its results.

Researchers, extension personnel, government and commercial propagators, and growers must all cooperate to ensure that these results are realized as rapidly as possible. While hard work and patience will be required, the Christmas tree industry will be rewarded with new standards of productivity, quality and profitability.

DEFINITION OF TREES OUTSIDE THE FOREST

Trees outside the forest are defined by default, as all trees excluded from the definition of forest and other wooded lands. Trees outside the forest are located on "other lands", mostly on farmlands and built-up areas, both in rural and urban areas. A large number of TOF consist of planted or domesticated trees. TOF include trees in agroforestry systems, orchards and small woodlots. They may grow in meadows, pastoral areas and on farms, or along rivers, canals and roadsides, or in towns, gardens and parks. Some of the land use systems include alley cropping and shifting cultivation, permanent tree cover crops (*e.g.* coffee, cocoa), windbreaks, hedgerows, home gardens and fruit-tree plantations.

Classification of trees outside the forest presents certain difficulties. There are existing classifications for agroforestry, but none applicable to all trees outside the forest. For practical reasons, the FRA 2000 definition of "forest" combines aspects of both land cover and land use. This approach creates difficulties not only for classification of forest, but also for classification of TOF. In a study on data-gathering on TOF in Latin America, where classification was primarily based on land use criteria, separating land use and land cover aspects was found to be a main source of misinterpretation. There was a possibility of confounding coffee plantations and trees in pasture with forest, given their high density.

This clearly shows some of the problems involved in establishing a simple and reliable *a posteriori* classification.In France, the National Forest Inventory (IFN) and Teruti Land Use Study have begun to attempt coordinating classifications of trees outside the forest. The objective is eventually to use the annual Teruti data to update the IFN ten-year data, with a single national nomenclature as a possible end result.

FUNCTIONS AND CHALLENGES

In industrialized countries, farmers list shade and shelter, soil protection and improvement of the landscape and rural environment as their main reasons for growing trees. In the tropics, farmers grow woody species for food security and subsistence. Trees outside the forest are a major source of food. Livestock fodder produced by TOF can be a matter of life and death in semi-arid or mountainous areas.

Fuelwood remains the prime source of energy in developing countries, representing up to 81 per cent of the wood harvest (FAO 1999). In contrast, in the industrialized countries, fuelwood accounts for less than 10 per cent of total fuel consumption. Very few studies have reported on overall fuelwood output from stands and single trees outside the forest, but agroforestry systems and orchards are known to provide a large part of the resource. Trees outside the forest have an important ecological role. Planted trees and shrubs in fields help to check runoff and erosion and control flooding, as well as helping to purify water and protect against wind. Trees lining rivers and streams help to maintain biodiversity, providing spawning beds for fish and shellfish and shade which reduces eutrophication.

The unique role of trees in soil protection and conservation, checking wind and water erosion and maintaining soil fertility is universally acknowledged. Also important are the cumulative benefits of trees on smallholdings to soil and water conservation, in particular in the larger context of mountain watershed management; their positive impact on climate; and their role in buffering the effects of desertification and drought.

RESULTS OF SELECTED STUDIES

In spite of the limits of data at the regional or global level, a number of local initiatives have been carried out. The approaches of the various studies differ in accordance with the purpose and scale of the analysis. Few studies use methods resembling the conventional forest inventory. Many studies rely on existing literature or estimates drawn from surveys and interviews. The quantification of products is often based on different parameters, such as estimates of global output, marketed output, observed or potential productivity or economic value. Thus the reliability of the results is uncertain.

The following are the results of some national initiatives that have assessed trees outside the forest. In Kerala, the most densely inhabited state of India, a study estimated that of the total annual production of 14.6 million cubic metres of wood in the state, about 83 per cent was from homesteads (house compounds and farmlands), 10 per cent from estates (plantations of rubber, cardamom, coffee and tea) and only about 7 per cent from forest areas 26.6 per cent of the state area is under forest cover. Trees outside the forest met about 90 per cent of the fuelwood requirements of the state. Fuel from coconut trees alone, including both wood and non-wood materials (pruned and fallen), constituted about 70 per cent of the total fuelwood supply.

A study in Haryana State in India, an intensively cultivated state with about 3.8 per cent of its area classified as forest land but only about 2 per cent under actual forest cover showed that farm forestry (trees along farm bonds and in small patches up to 0.1 ha) accounted for 41.2 per cent of the total growing stock of wood. Multiple tree rows along roads and canals accounted for 13 per cent and 9.6 per cent, respectively; village woodlots for 24 per cent; and block plantations of less than 0.1 ha for 10.6 per cent. In Morocco, where forest cover is less than 5 per cent of the land cover and other wooded lands only 7 per cent, nearly 20 per cent of the land may be occupied by trees outside the forest, namely as wooded pasture (84 per cent) and fruit-tree plantations (12 per cent) (Rosaceae, citrus, olives trees, palm trees, walnut trees, fig trees, almond trees).

Fruit production has an important place in the national economy. It is noteworthy that even when a forest is largely destroyed, the carob is one of the few species traditionally conserved, as it is highly appreciated by farmers for multiple purposes, providing both fodder and income from the sale of its fruit for export. However, there are no reliable data on the distribution and potential of this "forest" resource, which is of interest for farmers, herders, concessionaires and the government and distributed on agricultural and forest lands. In the Sudan, the National Forest Inventory has undertaken a national land use inventory to provide area and volume statistics for planning at the subnational and national levels. The inventory was designed to provide preliminary estimates regarding products other than the traditional fuelwood and timber, such as the amount of gum, fruit or nuts that can be collected and the distribution of non-wood species of interest.

In Costa Rica, the Tropical Agricultural Research and Higher Education Centre (CATIE), in collaboration with Freiburg University, Germany, is developing a regional methodology for Central America to assess tree resources outside the forest. A mix of satellite remote sensing, aerial photos and ground sampling is used to address the complexity of the resource and to allow dynamic monitoring of resources at the national and regional levels data-gathering on TOF in eight Latin American countries. None of these countries had established a database and the search for information was multisectoral. The statistics on land cover and land use gave some idea of the relative importance of trees outside the forest.

In Kenya, extensive tree planting on farmlands was promoted in the 1970s and 1980s, with land tenure security as a major incentive. There is an increasing trend of tree cover and species diversification on privately owned farms. Assuming that the present rate of increase in tree planting will continue, it was estimated that farms produced about 9.4 million cubic metres of wood in 2000 and will produce about 17.8 million cubic metres in 2020.

Their share of the total wood produced in the medium- and high-potential districts was projected to increase to 80 per cent in 2020. Indeed, while natural stands of trees have declined there has been a corresponding increase in tree planting in much of the densely populated plateaus of Kenya. As natural forests are reduced or become inaccessible, agroforestry systems help people to diversify production and income and to protect themselves from shortages of fuel and wood.

In Bangladesh, natural forest formations cover less than 6 per cent of the country and the population growth rate is extremely high. An inventory of homestead/village forests in the country indicated that trees outside the forest constitute a vital resource for local populations, providing food, fodder and fuelwood. The sampling method was based on dual village/household sampling with an agro-ecological and administrative sampling base. Rural Bangladesh was divided into six major regions considered as agro-ecological strata, each subdivided into *thanas* (administrative entities, subdistricts). The households making up the sampling units were chosen at random from a number of villages.

The inventory sampled data on palm trees and cane as well as trees, bamboo and thickets. The results, expressed per stratum and per inhabitant, provide volumetric data for fuelwood and sawnwood and species data for total amounts under and over 20 cm. This inventory was apparently the first to nationwide assessment of trees outside the classified forests in Bangladesh.

4

Genomics and Forest Biology

Forest biologists have developed strong justifications for why trees should be viewed as model systems in plant biology, including the obvious challenges in extrapolating findings from annual, herbaceous plants to organisms that are distinguished by perennial growth, large size, complex crown architecture, extensive secondary xylem, dormancy, and juvenile–mature phase changes. Similar justification has been used to argue why the genome of a tree should be sequenced. The U.S. Department of Energy (DOE), Office of Science, announced earlier this year plans to sequence the first tree genome, that of the black cottonwood (*Populus trichocarpa*).

In what has emerged as an exciting area in forest biology, two meetings recently convened at which advances in *Populus* genomics were featured: the International Poplar Symposium III, held August 26 to 29, 2002, in Uppsala, Sweden, and the*Populus* Functional Genomics Workshop, held August 30, 2002, in Umeå, Sweden. Together, the meetings attracted 200 scientists from 23 countries. Here, we summarize those portions of the symposium and workshop that highlight emerging tools, techniques, and research directions in *Populus* genomics.

SEQUENZCING THE POPLAR GENOME

Although efforts to identify *Populus* as a model tree began long before sequencing a tree genome was a possibility, the choice of poplar was ideal in that the genome size is small, <"550 Mbp. This is similar in size to the rice genome, only 4 times larger than the genome of Arabidopsis, yet 40 to 50 times smaller than the genome of pine.

Not surprisingly, a presentation by Gerald Tuskan from Oak Ridge National Laboratory (ORNL; Oak Ridge, TN) drew a large audience interested in learning more about efforts to sequence the *Populus* genome. According to preliminary plans, the DOE Joint Genome Institute, with funds from the DOE Office of Biological and Environmental Research, will provide a 3× draft sequence of the female black cottonwood clone Nisqually-1 in late 2002 and a second 3× draft in late 2003. The first 3× draft will be generated by a random shotgun approach. The second 3× draft will be based on a minimum tiling path that has been established from 90,000 BAC end sequences provided by Genome Canada (Vancouver, British Columbia). Together, these approaches will provide <"6× coverage of the genome. Alignment of assembled contigs will be based on information from physical and genetic maps. Annotation will take place using *Populus*-specific gene-finding models trained from full-length cDNA sequences provided by the Umeå Plant Science Center (UPSC; Umeå, Sweden), Genome Canada, and ORNL.

EST SEQUENCING

ESTs represent an informative tool for gene discovery. Stefan Jansson from UPSC reported on an extensive EST library being developed as part of the Swedish*Populus* Genome project, a joint collaboration between UPSC and the Genome Center at the Royal Institute of Technology in Stockholm. In the initial phase of this project, almost 5700 ESTs were developed for wood-forming tissues. Today, this resource has grown to >95,000 ESTs sequenced from 20 different cDNA libraries and from a range of tissues and developmental stages. Analyses indicate that these ESTs derive from perhaps 15,000 to 20,000 genes, a significant fraction of the 40,000 to 50,000 genes believed to be coded by the*Populus* genome. Basic Local Alignment Search Tool (BLAST) searches against sequenced ESTs are possible through the project database (PopulusDB) or data deposited in GenBank.

Although a functional classification of all 95,000 ESTs has not been completed, several subsets of the data have been analyzed. The distribution of genes in various functional categories for young poplar leaves and leaves harvested before visible signs of senescence. According to Jansson, young leaves devote one-third (36%) of their transcript pool to "energy," whereas older leaves have a high abundance of transcripts in categories such as "cell death and aging" and "protein destination," which includes functions related to proteolytic degradation. Furthermore, the fraction of sequences with little similarity to known genes (BLAST score of <100) was almost twice as high in autumn leaves as it was in young leaves.

Most of the ESTs sequenced to date have been for hybrid aspen (*P. tremula* × *P. tremuloides*), European aspen (*P. tremula*), and *P. trichocarpa*. Chung-Jui Tsai from Michigan Technological University (Houghton) reported that >12,000 ESTs also have been sequenced from various tissues of quaking aspen (*P. tremuloides*). Based on initial comparisons, gene sequences among the different species show high (>95%) similarities. Sequence similarities also exist between poplar ESTs and ESTs from other tree genera, including *Salix* (i.e., willow). Rishikesh Bhalerao from UPSC described a *Salix* EST sequencing program in which a cDNA library has been prepared from stem tissues. To date, <"2000 *Salix* ESTs have been sequenced and compared with ESTs from Arabidopsis and poplar. Analyses indicate a moderate degree of similarity with Arabidopsis but a high degree of sequence conservation between *Populus* and *Salix*. Such conservation suggests that resources being developed for *Populus* (e.g., microarrays) also can be used in related genera.

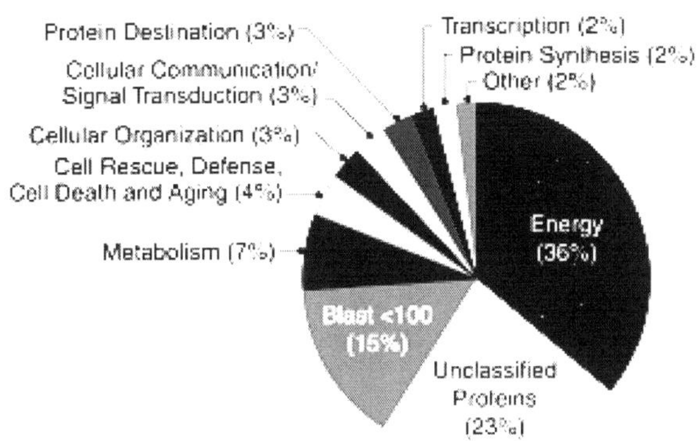

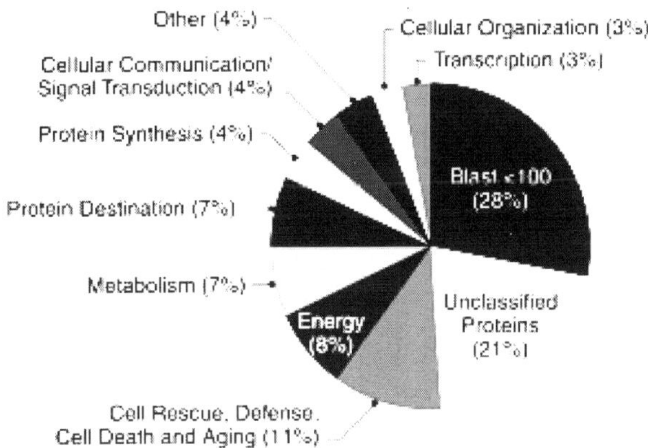

Fig. Functional Distribution of Genes According to a Modified MIPS (Munich Information Center for Protein Sequences) Classification Scheme of 4842 ESTs from Young *Populus* Leaves and 5128 ESTs from Leaves Collected in Autumn.

GENE EXPRESSION MICROARRAYS

Transcript profiling can reveal patterns of gene expression during developmentally regulated events such as wood formation and leaf expansion and maturation. Göran Sandberg from UPSC described how early poplar microarrays were used to study developmental processes involved in wood formation. In this case, microarrays helped characterize how gene expression varied throughout cell division, expansion, secondary wall formation, lignification, and cell death. These studies used a microarray spotted with 2995 ESTs unique to hybrid aspen.

Preparation of a higher density *Populus* microarray based on the 95,000-EST library described above has been initiated. As reported by Peter Nilsson from the Royal Institute of Technology in Stockholm, the Swedish *Populus* Genome project has produced a spotted EST microarray containing >13,000 clones. This array is based on the unigene set extracted after sequencing 35,000 clones. Swedish researchers are using these microarrays to study many processes in *Populus*, and access to the arrays will be provided on a collaborative basis. One such collaboration was described by Gail Taylor (Southampton University, UK), who presented data on global gene expression after long-term exposure to increased CO_2 concentrations of poplar trees grown as part of the PopFACE (free air carbon dioxide enrichment) experiment in central Italy. Approximately 1500 array ESTs showed enhanced expression, whereas 1000 ESTs showed reduced expression, at increased CO_2 concentrations. Taylor emphasized that these studies eventually should provide much-needed insights into the long-term adaptation of trees to future climate conditions. The challenge, of course, will be to determine how changes in gene expression are related to altered biochemical and physiological function and, ultimately, to tree growth.

Transgenic Lines and Functional Genomics

The creation of transgenic lines with enhanced or reduced levels of gene expression is the most straightforward way to determine gene function. Rishikesh Bhalerao described an ambitious program in which a commercial partner, SweTree Genomics, will use RNA interference to knock out 2000 genes involved in wood formation. A high-throughput recombinational cloning system will allow the necessary constructs to be generated within 2 years and the production of regenerated transgenic plants within 48 months. The group has devised a tiered phenotyping program that, as reported by Björn Sundberg from UPSC, will identify transformed lines with altered physical and chemical properties, thus highlighting economically and biologically important processes involved in wood formation.

Richard Meilan (Oregon State University, Corvallis) described activation tagging as an approach to the creation of dominant mutations in trees. Activation tagging is a method whereby an enhancer element is inserted randomly into the genome, enhancing the expression of a nearby gene. Meilan and Oregon State University colleagues Amy Brunner, Victor Busov, and Steve Strauss have used this technique to produce a large population of transformed individuals. Nine of these plants showed a visibly altered phenotype. One mutant exhibiting stunted growth, thick stems, and dark leaves was identified as having an open reading frame with high nucleotide similarity to a gibberellic acid 2-oxidase gene. The Oregon State University group also is working with Andrew Groover (U.S. Forest Service, Davis, CA) to identify genes based on expression patterns using enhancer-trap and gene-trap insertion lines. In this project, T-DNAs carrying the β-glucuronidase reporter gene are introduced randomly into the genome, and insertion results in a pattern of β-gluc-uronidase expression that mimics the normal expression patterns of interrupted genes. These techniques are being used to identify genes with preferential expression in leaf vascular tissues, wood-forming tissues, roots, and adventitious root primordia.

In addition to T-DNA–based methods, an alternative transposable tagging system is being explored by Sandeep Kumar and Matthias Fladung from the Institute for Forest Genetics and Forest Tree Breeding (Grosshansdorf, Germany) for use in generating mosaics of insertional and activation mutants within a single plant. The system combines the dominant *rolC* marker gene from *Agrobacterium rhizogenes* and the *Ac* transposon from maize. Excision of the *Ac* element is indicated by pale green leaf sectors caused by *rolC* expression, thus visually marking cells in which the *Ac* element has excised native genes upon reinsertion. Plants then can be regenerated from the affected cells in tissue culture. A comparison of *Ac*/T-DNA insertional data indicated that rates of *Ac* insertion in coding regions were twofold greater than for T-DNA–based systems.

It is anticipated that the number of transgenic lines produced by the methods described above will eventually exceed the capacity for their maintenance and characterization. Chung-Jui Tsai and colleagues are investigating the feasibility of using a cryonics-based preservation system for the long-term storage of transgenic germplasm.

METABOLIC CHARACTERIZATION OF PHENOTYPES

Because visual differences among transformed individuals are expected in <1% of a transgenic population, methods to characterize the metabolic phenotypes of trees possessing gain- or loss-of-function mutations must be developed. Consistent with the terminology of Fiehn (2002), metabolic profiling is being explored in *Populus* by Chung-Jui Tsai and Tim Tschaplinski from ORNL, and metabolomics is being explored by Thomas Moritz from UPSC. Each group is using similar sample preparation and derivatization protocols, followed by either quadrupole gas chromatography–mass spectrometry or fast gas chromatography–time-of-flight mass spectrometry. Few results were presented, although everyone agreed that the large amount of data generated from these analyses would dictate compromises between speed of analysis and number of metabolites detected and quantified. Multiple online fractionation techniques are being considered, and additional analyses involving liquid chromatography–mass spectrometry are being developed to detect metabolites that are difficult to analyze by gas chromatography–mass spectrometry.

Given that many of these techniques are just being developed, all speakers agreed that various analytical and computational challenges would be encountered in this area of investigation. Thomas Moritz illustrated the complexities that will have to be overcome in *Populus* metabolomics by discussing how fragmentation patterns for hundreds of compounds might be detected and analyzed statistically using principal-component analysis and partial least-squares projections to latent structures. Such analyses would eventually link data obtained with expression arrays to consequences in biochemical pathways, and thus to function. Tim Tschaplinski pointed out that there are many unique low molecular mass compounds in *Populus* and suggested that the establishment of standardized mass spectrometry libraries would aid greatly in the development of subsequent deconvolution strategies.

Formation of an International Genome Consortium

One concern that was raised during the International Poplar Symposium III, and again at the *Populus* Functional Genomics Workshop, was the fact that the genomics community in forestry is small compared with other plant science communities (e.g., rice and Arabidopsis). As such, participants emphasized that groups conducting research in *Populus* should strive to organize their scientific agendas to complement, not compete with, one another. This coordination would enable groups to avoid unnecessary duplication and allow limited resources to be focused on specific problems that broadly challenge the community.

Returning again to the podium, Gerald Tuskan proposed the formation of an International *Populus* Genome Consortium, the purpose of which would be to develop and guide postsequencing activities in poplar. Foremost among the goals highlighted for the consortium would be the development of a *Populus* science plan. Teams would be formed to examine genetic and genomic resources currently available to *Populus* researchers, identify areas in which tools, techniques, and additional resources must be developed, and assess applications and opportunities for future research associated with the completion of the *Populus* genome sequence. Applications would emphasize tree growth and development in the context of general poplar culture, basic science investigations, bio-based products and energy, carbon sequestration, and forest responses to changes in physical and chemical climates. The formation of an inter-national consortium and the development of a science plan were endorsed strongly by symposium and workshop participants.

THE GREAT DIVERSITY FOUND IN FOREST TREES

There are an estimated 60,000 tree species on earth, and approximately 30 of the 49 plant orders contain tree species. Clearly, the tree phenotype has evolved many times in plants. The diversity of plant structures, development, life history, environments occupied and so on in trees is nearly as broad as higher plants in general, but trees share the common characteristic that all are perennial and many are very long lived. Because of the sessile nature of plants, each tree must survive and reproduce in a specific environment over the seasonal cycles of its lifetime. This tight association between individual genotypes and their environment provides a powerful research setting, just as it has driven the evolution of a plethora of uniquely arboreal adaptations. Understanding these evolutionary strategies is a long-standing area of study of tree biologists, with many broader biological implications.

Completed and current genome-sequencing projects in forest trees are limited to about 25 species from just 4 of more than 100 families: Pinaceae (pines, spruces and firs), Salicaceae (poplars and willows), Myrtaceae (eucalyptus) and Fagaceae (oaks, chestnuts and beeches). Large-scale sequencing projects such as the 1000 Human Genomes, 1000 Plant Genomes (1KP) or the 5000 Insect Genome (i5k) projects have not yet been proposed for forest trees.

RAPIDLY DEVELOPING GENOMIC RESOURCES IN FOREST TREES

Genome resources are developing rapidly in forest trees in spite of the challenges associated with working with large, long-lived organisms and sometimes very large genomes [2]. Complete genome sequencing, however, has been slow to advance in forest trees owing to funding limitations and the large size of conifer genomes. Black cottonwood (*Populus trichocarpa* Torr. & Gray) was the first forest tree genome to be sequenced by the US Department of Energy Joint Genome Institute (DOE/JGI) [6] (Table 1). Black cottonwood has a relatively small genome (450 Mb) and is a target feedstock species for cellulosic ethanol production, and thus fits into the DOE/JGI priority of sequencing bioenergy feedstock species. The genus *Populus* has 30+ species (aspens and cottonwoods) with genome sizes of approximately 500 Mb. Several species are being sequenced by DOE/JGI, and other groups around the world, and it seems likely that all members of the genus will soon have a genome sequence (Table 1). The next forest tree to be sequenced was the flooded gum (*Eucalyptus grandis* BRASUZ1, which is a member of the Myrtaceae family), again by DOE/JGI. Eucalyptus species and their hybrids are important commercial species grown in their native Australia and many regions throughout the southern hemisphere. Several more eucalyptus species are being sequenced, each with relatively small genomes (500 Mb), but it will probably take many years before all 700+ members of this genus are completed. Several members of the Fagaceae family are now being sequenced (Table 1). Members of this group include the oaks, beeches and chestnuts, with genome sizes less than 1 Gb.

The gymnosperm forest trees (such as the conifers) were the last to enter the world of genome sequencing. This was entirely due to their very large genomes (10 Gb and greater) as they are extremely important economically and ecologically, and phylogenetically they represent the ancient sister lineage to that of angiosperm species. Genome resources needed to support a sequencing project were reasonably well developed, but it was not until the introduction of next-generation sequencing (NGS) technologies that sequencing conifer genomes became tractable. Currently, there are at least ten conifer (Pinaceae) genome-sequencing projects under way. Aside from reference genome sequencing in forest trees, there is significant activity in transcriptome sequencing and resequencing for polymorphism discovery (Tables 2 and 3). We have only listed the transcriptome and resequencing projects in Table 1 that are associated with a species that has an active genome-sequencing project.

THE OPPORTUNITY FOR COMPARATIVE-GENOMIC APPROACHES IN FOREST TREES

The power of comparative-genomic approaches for understanding function in an evolutionary framework is well established. Comparative genomics can be applied to sequence data (nucleotide and protein) at the level of individual genes or genome-wide. Genome-wide approaches provide insight into both chromosome evolution and the diversification of biological functions and interactions. Understanding of gene function in forest tree species is challenged by the lack of standard reverse-genetic tools routinely used in other systems - for example, standard marker stocks, facile transformation and regeneration - and by the long generation times. Thus, comparative genomics becomes the more powerful approach to understanding gene function in trees.

Comparative genomics requires not only data availability but also cyber-infrastructure to support exchange and analysis. The TreeGenes database is the most comprehensive resource for comparative-genomic analyses in forest trees. Several smaller databases have been created to facilitate collaborations, including: Fagaceae genomics web, hardwoodgenomics.org, Quercus portal, PineDB, ConiferGDB, EuroPineDB, PopulusDB, PoplarDB, EucalyptusDB and Eucanext (Tables 1, 2, and 3). These resources vary greatly in their scope, relevance and integration. Some are static and archival, whereas others focus on current sequence content for a specific species or a small number of related species. This results in overlapping and conflicting data among repositories. In addition, each database uses its own custom interfaces and back-end database technology to serve sequence to the user.

The US National Science Foundation funding for large-scale infrastructure projects, such as iPlant, is leading efforts aimed towards centralizing resources for research communities [15]. Without centralized resources, researchers are forced to employ inefficient data-mining methods through queries of independently maintained databases or inconsistently formatted supplemental files on journal websites. Specific areas of interest for the forest tree genomic community include the ability to connect sequence, genotype and phenotype to individual, geo-referenced trees. This type of integration can only be achieved through web services that allow disparate resources to communicate in ways that are transparent to the user[16]. With the recent increase of genome sequences available for many of these species, there is a need to facilitate community-level annotation and research support.

The need for a Better-Developed Open-Access Culture

The Human Genome Project established a culture of open access and data sharing in genomics research for both humans and animal models that has been extended to many other species, including *Arabidopsis*, rat, cow, dog, rice, maize and more than 500 other eukaryotes. Beginning in the late 1990s, these large-scale projects released data very rapidly to the scientific community, often years before publication. This rapid release of data with few restrictions has allowed thousands of scientists to begin work on specific genes and gene families, and on functional studies, long before the genome papers have appeared. One of the driving motivations for this culture, and the reason that many scientists support it, is that large-scale sequencing can be done most efficiently when centers that have expertise in sequencing technology take the lead. With all the sequencing concentrated, the body of data needs to be shared freely in order to get it in the hands of the widely distributed experts. This open-access culture has dramatically accelerated scientific progress in biological research.

The Path to Success Avoids Delays

Careful inspection of Table 1 reveals that forest tree genome projects are very slow to release sequence data into the public domain. Once a project is finished and submitted for publication, a draft genome becomes available - for example, the poplar genome was released and published in 2006. However, pre-publication releases are infrequent, exceptions being the PineRefSeq project that has made three releases and the SMarTForest project that has made one (Table 1). This is unfortunate because good-quality sequence contigs and scaffolds could be made available years before publication, delivering an extremely important resource to the community. This delay can be understood from privately financed projects seeking commercial advantages, but nearly all the projects listed in Table 1 are financed by public funds whose stated mission is advancing science and development of community resources. Publication rights are easily protected by data-use policy statements such as the Ft Lauderdale [17] and Toronto agreements [18], but unfortunately these conventions are not often used and data access is restricted by password-protected websites (Tables 1, 2, and 3). We hope the opinion offered here will lead to a discussion in the forest tree community, to a more open-access culture and thus to a more vibrant and rapidly advancing research area.

PROPOSAL TO SEQUENCE POPULUS GENOME

Forest trees are the dominant life form in many ecosystems and contain greater than 90% of the Earth's terrestrial biomass. Managed and unmanaged forests throughout the United States, and indeed the world, provide recreational and environmental benefits such as carbon sequestration, renewable energy supplies, watershed protection, improved air quality, biodiversity and habitat for endangered species. However, despite the importance of forest trees for natural ecosystems and the world economy, little is known about the biology of forest trees in comparison with the detailed information available for crop plants and model organisms such as*Arabidopsis*. As a result, the forest science community would derive much benefit from a comprehensive genomics research program, because traditional genetic approaches in forestry are limited by the large size, long generation interval, and outcrossing mating system of most trees. It is, therefore, of considerable importance that a forest tree genome be sequenced, so that forest tree biologists will have the necessary resources to begin a large scale, thorough analysis of genes that produce traits useful in the pursuit of basic science questions, to foster the development of improved plant materials for the forest products industry, and to ultimately select novel phenotypes that could be used to address questions related to the energy-related mission of the Department of Energy.

The genus *Populus* (including poplars, cottonwoods and aspens) is especially well suited to serve as the model genome for trees because of the following reasons:

- Small genome size; the haploid genome size of *Populus* is only ca. 550 Mbp, similar to rice, only 4X larger than *Arabidopsis* and 40X smaller than pine.
- 30^+ species worldwide; all are diploids ($2n = 38$) that can be easily crossed, hybrids are fertile and advanced generation pedigrees are available.
- Rapid juvenile growth, allowing meaningful measures of important traits to be taken within a few years after establishment of genetic trials.
- Ease of clonal propagation, allowing manipulations to be evaluated across common genetic backgrounds, destructive sampling (e.g., wood quality assessment) without loss of the genotype, replication of experiments across time and space, and genetic stocks to be archived in clonal nurseries to be shared with researchers around the world.
- High-throughput transformation and regeneration; transgenic trees can be produced allowing detection and characterization of gene function. *Populus* is unique among tree genera in this regard.
- Extensive genetic maps have been constructed in *Populus*, including the initial identification of QTLs (i.e., DNA markers associated with) a wide variety of traits.
- Existing collaborative network of scientists around the world who have used poplars as a model organism to study forest tree

morphology, physiology, biochemistry, ecology, genetics, and molecular biology.

THE IMPORTANCE OF THE GENOME TO DOE MISSION AND STATED GOALS

The Department of Energy, through its demonstrated success in the Human Genome Project and various microbial sequencing activities, is in a strong position to champion a *Populus* genome initiative. Such an activity focused on a woody perennial would bring together strengths in molecular biology, computational biology, bioinfomatics, global climate change, carbon management, bioremediation and thereby, would draw upon the considerable expertise provided by multiple programs within DOE's Office of Science. In terms of research sponsored by the Office of Biological and Environmental Research (BER), *Populus* species including cottonwood, hybrid poplar and aspen are already being used in many activities ranging from carbon sequestration research, to free-air CO_2 enrichment (FACE) studies, and to the development of fast-growing trees as a renewable bioenergy resource (EE/RE). Furthermore, a sequencing effort in trees could conceptually be applied to areas of phytoremediation, whereby trees such as poplars could be used to remediate hazardous waste sites (http://www.anl.gov/OPA/Frontiers2000/d3ee.html). Clearly, the information derived from a genome sequencing effort would benefit projects within DOE and open the doors to countless other opportunities to use woody plants in the pursuit of questions of interest to the energy-mission of DOE.

The size and interest of the research community

The Plant and Animal Genome Meeting, a session was organized to address the possibility of soliciting support for sequencing the genome of a forest tree. Support among attendees was high and the organizers of that session established a website for promoting the establishment of an international collaborative, public effort to determine the complete DNA sequence of the poplar (*Populus*) genome.

It was felt that before the forest science community could proceed to solicit the political support necessary to begin a poplar genome sequencing project, it would be vital that forest tree biologists express their enthusiasm for such a project. So far, over 70 scientists have indicated via the website that they acknowledge that a poplar genome sequencing effort should be a top priority of forest genomics research and have expressed their desire to see this activity proceed as quickly as possible toward the ultimate goal of determining the complete sequence of the *Populus* genome.

Resources available to complement the sequence

The genus *Populus*, one of two genera in the Family Salicaceae, first occurred in the fossil record around 56 m.y.b.p. Today, the genus consists of five sections, 30 to 40 species and is circumpolar in the Northern Hemisphere. There are approximately eight species indigenous to North America, with four of those having commercial importance. All members of the *Populus* genus have a genome contained on 19 nearly identical, metacentric chromosomes, a nuclear content of 2C = 1.2 pg and an estimated genetic map length of ca. 2500 cM. The majority of the species are constitutively dioecious, having male and female members of a species. Interspecific hybridization is readily achievable among members within a single section and among members of certain alternate sections. Selection and hybridization began among the aspens in the middle 1950s, followed by selection and breeding in cottonwoods in the 1960s. Heterosis or hybrid vigor is common among interspecific hybrids. Relatively large multi-generation pedigrees exist for those species and their hybrids that have commercial value.

The most commonly produced pedigrees are from *P. trichocarpa* x *P. deltoides, P. deltoides* x *P. nigra, P. grandidentata* x *P. alba* and *P. tremuloides* x *P. tremula*. These species provide a logical choice for a genome sequencing effort. The largest F_1 pedigree [2000^+ progeny] is between a *P. trichocarpa* female '383-2499' and a *P. deltoides* male 'ILL129'. The female, '383-2499', has provided the nuclear DNA for the construction of a 10x BAC library that has been distributed around the world. This BAC library was originally constructed by scientists at Texas A&M University.

Preliminary efforts to clones genes from alternate genotypes suggest that homology is high among all members of the genus [$90\%^+$] and that *Populus* contains multiple, typically two, copies of single genes found in other plant species such as *Arabidopsis*. As such, we recommend that the female clone '383-2499' be used in the genomic sequence effort for all *Populus*. This *P. trichocarpa* or black cottonwood female was originally collected along the Nisqually River in Washington, and is replicated in field and nursery plantings presently maintained by Dr. Toby Bradshaw and colleagues, University of Washington and Washington State University.

Indications on how the genome sequence would be used

Once a draft sequence is available for the *Populus* genome, steps would be taken to annotate the completed genome. This work would be done in collaboration with JGI and ORNL, with the participation of the forest genetics community. We are fortunate to have access to Dendrome, a web-based collection of forest tree genome databases and other forest genetic information resources for the international forest genetics community.

RANDOM FOREST FOR GENE SELECTION AND MICROARRAY DATA CLASSIFICATION

A random forest method has been selected to perform both gene selection and classification of the microarray data. In this embedded method, the selection of smallest possible sets of genes with lowest error rates is the key factor in achieving highest classification accuracy. Hence, improved gene selection method using random forest has been proposed to obtain the smallest subset of genes as well as biggest subset of genes prior to classification. The option for biggest subset selection is done to assist researchers who intend to use the informative genes for further research. Enhanced random forest gene selection has performed better in terms of selecting the smallest subset as well as biggest subset of informative genes with lowest out of bag error rates through gene selection. Furthermore, the classification performed on the selected subset of genes using random forest has lead to lower prediction error rates compared to existing method and other similar available methods.

Through various biological experiments conducted worldwide, large datasets of information has been increasing rapidly and more analysis is conducted each day to sort out the puzzle. Since there are many separate methods available for performing gene selection as well as classification, finding similar approach for both, has been of interest to many researchers. Gene selection focuses at identifying a small subset of informative genes from the initial data in order to obtain high predictive accuracy for classification. Gene selection can be considered as a combinatorial search problem and therefore can be suitably handled with optimization methods. Besides that, gene selection plays an important role preceding to tissue classification, as only important and related genes are selected for the classification. The main reason to perform gene selection is to identify a small subset of informative genes from the initial data before classification in order to obtain higher prediction accuracy. Many researchers use single variable rankings of the gene relevance and random thresholds to select the number of genes, which can only be applied to two class problems. Random forest can be used for problems arising from more than two classes (multi class) as stated by Díaz-Uriarte R & Alvarez de Andrés (2006).

Classification is carried out to correctly classify the testing samples according to the class. Therefore, performing gene selection antecedent to classification would severely improve the prediction accuracy of the microarray data. Random forest is an ensemble classifier which uses recursive partitioning to generate many trees and then combine the result. Using a bagging technique first proposed by Breiman (1996), each tree is independently constructed using a bootstrap sample of the data. Classification generates gene expression profiles which can discriminate between different known cell types or conditions as described by Lee *et al.* (2004). A classification problem is said to be binary in the event when there are only two class labels present and a classification problem is said to be a multiclass classification problem if there are at least three class labels. An enhanced version of gene selection using random forest is proposed to improve the gene selection as well as classification in order to achieve higher prediction accuracy. The proposed idea is to select the smallest subset of genes with the lowest out of bag (OOB) error rates for classification. Besides that, the selection of biggest subset of genes with the lowest OOB error rates is also available to further improve the classification accuracy. Both options are provided as the gene selection technique is designed to suit the clinical or research application and it is not restricted to any particular microarray dataset. Apart from that, the option for setting the minimum number of genes to be selected is added to further improve the functionality of the gene selection method. Therefore, the minimum number of genes required can be set for gene selection process.

METHODOLOGY

Few improvements have been made to the existing random forest gene selection, which includes automated dataset input that simplifies the task of loading and processing of the dataset to an appropriate format so that it can be used in this software. Furthermore, the gene selection technique is improved by focusing on smallest subset of genes while taking into account lowest OOB error rates as well as biggest subset of genes with lowest OOB error rates that could increase the prediction accuracy. Besides that, additional functionalities are added to suite different research outcome and clinical application such as the range of the minimum required genes to be selected as a subset. Integration of the different approaches into a single function with parameters as an option allows greater usability while maintaining the computation time required.

Selection of Smallest Subset of Genes with Lowest OOB Error Rates

The existing method performs gene selection based on random forest to select smallest subset of genes while compromising on the out of bag (OOB) error rates. The subset of genes is usually small but the OOB error rates are not the lowest out of all the possible selection through backward elimination. Therefore, enhancement has been made to improve the prediction accuracy by selecting the smallest subset with the lowest OOB error rates. Hence, lower prediction error rates can be achieved for classification of the samples. This technique is implemented in the random forest gene selection method and shown in Figure 3 (see supplementary material) under supplementary section. During each subset selection based on backward elimination, the mean OOB error rate and standard deviation OOB error rate are tracked at every loop as the less informative genes are removed gradually. Once the loop terminates the subset with the smallest number of variables and lowest OOB error rates are selected for classification. The subset of genes is located based on the last iteration with the smallest OOB error rates. During the backward elimination process, the number of selected variables decreases as the iteration increases.

Selection of Biggest Subset of Genes with Lowest OOB Error Rates

Another method for improving the prediction error rates is by selecting the biggest subset with the lowest OOB error rates. This is due to the fact that any two or more subsets with different number of selected variables with same lowest error rates indicates that the informative genes level are the same, but the contribution of each genes towards the prediction accuracy is not the same. So, having more informative genes can increase the classification accuracy of the sample. The technique applied for the selection of biggest subset of genes with the lowest OOB error rates are similar to the smallest subset of genes with the lowest OOB error rates, except that the selection is done by picking the first subset with the lowest OOB error rates from all the selected subset which has the lowest error rates. If there is more than one subset with lowest OOB error rates, the selection of the subset is done by selecting the one with highest number of variables for this method.

The detailed process flow for this method can be seen in the Figure 4 (see supplementary material). This technique is implemented to assist researches that require filtration of genes for reducing the size of microarray dataset while making sure that the numbers of informative genes are high. This is achieved by eliminating unwanted genes as low as possible while achieving highest accuracy in prediction. Further enhancement is made to the existing random forest gene selection process by adding an extra functionality for specifying the minimum number of genes to be selected in the gene selection process that is included into the classification of the samples. This option allows flexibility of the program to suite the clinical research requirements as well as other application requirement based on the number of genes needed to be considered for classification. The selected minimum values are used during the backward elimination process which takes place in determining the best subset of genes based on out of bag (OOB) error rates.

Performance Measurement

For gene selection using random forest, backward elimination using OOB error rates is used as the final set of genes is selected based on the lowest out of bag (OOB) error rates as random forest returns a measure of error rate based on the out-of-bag cases for each fitted tree. The classification performance of the microarray data using random forest is measured using.632 bootstrap methods. In this method, the prediction error rates obtained is used to compare the performance of the random forest in classification where lower error rates means higher prediction accuracy. In the .632 bootstrap, accuracy is estimated as followed. Given a dataset of size n, a bootstrap sample is created by sampling n instances uniformly from the data (with replacement). Since the dataset is sampled with replacement, the probability of any given instance not being chosen after n samples is given in the supplementary material.

The expected number of distinct instances from the original dataset appearing in the test set is thus 0.632. The accuracy estimate is derived by using the bootstrap sample for training and the rest of the instances for testing. Given a number *b*, the number of bootstrap samples, let $c0_i$ be the accuracy estimate for bootstrap sample *i*. The .632 bootstrap estimates are defined as given in the supplementary material. The assessment method used has been able to populate and list the overall performance of the algorithm with other similar algorithms and techniques through prediction error rates calculation comparison.

Results & Discussion

In this section, the full result of all the options used is compared. In Figure 1, the result for each dataset is plotted against the accuracy, therefore the higher the values the lower is the error rates. Based on the Figure 1, the enhanced random forest gene selection performs better compared to standard method. Though, different options have different effects to the datasets being tested. Most of the datasets tested showed larger improvement in terms of accuracy achieved for classification when the subset of genes selected is larger. The detailed information regarding the datasets has been tabulated in Table 1 (see supplementary material). However, some datasets with smaller subset of genes outperformed the larger subset of genes. This could be due to the effect of the informative genes, as more informative genes contribute to better classification accuracy. For the Leukemia dataset, either the selection of biggest subset of genes or limiting the range of the minimum number of genes to be selected in a particular subset has reduced the prediction accuracy. This is due to the fact that low number of informative genes contributes less to the overall classification accuracy. The highest accuracy achieved for this dataset is by selecting smallest

GENE SELECTION AND CLASSIFICATION OF MICROARRAY DATA USING RANDOM FOREST

Selection of relevant genes for sample classification (e.g., to differentiate between patients with and without cancer) is a common task in most gene expression studies e.g.,. When facing gene selection problems, biomedical researchers often show interest in one of the following objectives:

1. To identify relevant genes for subsequent research; this involves obtaining a (probably large) set of genes that are related to the outcome of interest, and this set should include genes even if they perform similar functions and are highly correlated.
2. To identify small sets of genes that could be used for diagnostic purposes in clinical practice; this involves obtaining the smallest possible set of genes that can still achieve good predictive performance (thus, "redundant" genes should not be selected).

We will focus here on the second objective. Most gene selection approaches in class prediction problems combine ranking genes (e.g., using an *F*-ratio or a Wilcoxon statistic) with a specific classifier (e.g., discriminant analysis, nearest neighbor). Selecting an optimal number of features to use for classification is a complicated task, although some preliminary guidelines, based on simulation studies by [4], are available. Frequently an arbitrary decision as to the number of genes to retain is made e.g., keep the 50 best ranked genes and use them with a linear discriminant analysis as in; keep the best 150 genes as in. This approach, although it can be appropriate when the only objective is to classify samples, is not the most appropriate if the objective is to obtain the smaller possible sets of genes that will allow good predictive performance. Another common approach, with many variants e.g., is to repeatedly apply the same classifier over progressively smaller sets of genes (where we exclude genes based either on the ranking statistic or on the effect of the elimination of a gene on error rate) until a satisfactory solution is achieved (often the smallest error rate over all sets of genes tried). A potential problem of this second approach, if the elimination is based on univariate rankings, is that the ranking of a gene is computed in isolation from all other genes, or at most in combinations of pairs of genes, and without any direct relation to the classification algorithm that will later be used to obtain the class predictions. Finally, the problem of gene selection is generally regarded as much more problematic in multi-class situations (where there are three or more classes to be differentiated), as evidence by recent papers in this area e.g.,. Therefore, classification algorithms that directly provide measures of variable importance (related to the relevance of the variable in the classification) are of great interest for gene selection, specially if the classification algorithm itself presents features that make it well suited for the types of problems frequently faced with microarray data. Random forest is one such algorithm.

Random forest is an algorithm for classification developed by Leo Breiman that uses an ensemble of classification trees [14-16]. Each of the classification trees is built using a bootstrap sample of the data, and at each split the candidate set of variables is a random subset of the variables. Thus, random forest uses both bagging (bootstrap aggregation), a successful approach for combining unstable learners [16,17], and random variable selection for tree building. Each tree is unpruned (grown fully), so as to obtain low-bias trees; at the same time, bagging and random variable selection result in low correlation of the individual trees. The algorithm yields an ensemble that can achieve both low bias and low variance (from averaging over a large ensemble of low-bias, high-variance but low correlation trees). Random forest has excellent performance in classification tasks, comparable to support vector machines. Although random forest is not widely used in the microarray literature, it has several characteristics that make it ideal for these data sets:

- Can be used when there are many more variables than observations.
- Can be used both for two-class and multi-class problems of more than two classes.
- Has good predictive performance even when most predictive variables are noise, and therefore it does not require a pre-selection of genes (i.e., "shows strong robustness with respect to large feature sets",*sensu*).

- Does not overfit.
- Can handle a mixture of categorical and continuous predictors.
- Incorporates interactions among predictor variables.
- The output is invariant to monotone transformations of the predictors.
- There are high quality and free implementations: the original Fortran code from L. Breiman and A. Cutler, and an R package from A. Liaw and M. Wiener.
- Returns measures of variable (gene) importance.
- There is little need to fine-tune parameters to achieve excellent performance. The most important parameter to choose is *mtry*, the number of input variables tried at each split, but it has been reported that the default value is often a good choice. In addition, the user needs to decide how many trees to grow for each forest (*ntree*) as well as the minimum size of the terminal nodes (*nodesize*). These three parameters will be thoroughly examined in this paper.

Given these promising features, it is important to understand the performance of random forest compared to alternative state-of-the-art prediction methods with microarray data, as well as the effects of changes in the parameters of random forest. In this paper we present, as necessary background for the main topic of the paper (gene selection), the first through examination of these issues, including evaluating the effects of *mtry*, *ntree* and *nodesize* on error rate using nine real microarray data sets and simulated data.

The main question addressed in this paper is gene selection using random forest. A few authors have previously used variable selection with random forest and use filtering approaches and, thus, do not take advantage of the measures of variable importance returned by random forest as part of the algorithm. Svetnik, Liaw, Tong and Wang propose a method that is somewhat similar to our approach. The main difference is that first find the "best" dimension (p) of the model, and then choose the p most important variables. This is a sound strategy when the objective is to build accurate predictors, without any regards for model interpretability. But this might not be the most appropriate for our purposes as it shifts the emphasis away from selection of specific genes, and in genomic studies the identity of the selected genes is relevant (e.g., to understand molecular pathways or to find targets for drug development).

The last issue addressed in this paper is the multiplicity (or lack of uniqueness or lack of stability) problem. Variable selection with microarray data can lead to many solutions that are equally good from the point of view of prediction rates, but that share few common genes. This multiplicity problem has been emphasized by and and recent examples are shown in Although multiplicity of results is not a problem when the only objective of our method is prediction, it casts serious doubts on the biological interpretability of the results. Unfortunately most "methods papers" in bioinformatics do not evaluate the stability of the results obtained, leading to a false sense of trust on the biological interpretability of the output obtained. Our paper presents a through and critical evaluation of the stability of the lists of selected genes with the proposed (and two competing) methods.

In this paper we present the first comprehensive evaluation of random forest for classification problems with microarray data, including an assessment of the effects of changes in its parameters and we show it to be an excellent performer even in multi-class problems, and without any need to fine-tune parameters or pre-select relevant genes. We then propose a new method for gene selection in classification problems (for both two-class and multi-class problems) that uses random forest; the main advantage of this method is that it returns very small sets of genes that retain a high predictive accuracy, and is competitive with existing methods of gene selection.

EVALUATION OF PERFORMANCE AND COMPARISONS WITH ALTERNATIVE APPROACHES

We have used both simulated and real microarray data sets to evaluate the variable selection procedure. For the real data sets, original reference paper and main features and further details are provided in the supplementary material. To evaluate if the proposed procedure can recover the signal in the data and can eliminate redundant genes, we need to use simulated data, so that we know exactly which genes are relevant. Details on the simulated data are provided in the methods and in the supplementary material.

Estimation of error rates

To estimate the prediction error rate of all methods we have used the .632+ bootstrap method. The .632+ bootstrap method uses a weighted average of the resubstitution error (the error when a classifier is applied to the training data) and the error on samples not used to train the predictor (the "leave-one-out" bootstrap error); this average is weighted by a quantity that reflects the amount of overfitting. It must be emphasized that the error rate used when performing variable selection is not what we report in as prediction error rate. To calculate the prediction error rate as reported, for example, the .632+ bootstrap method is applied to the complete procedure, and thus the samples used to compute the leave-one-out bootstrap error used in the .632+ method are samples that are not used when fitting the random forest, or carrying out variable selection. The .632+ bootstrap method was also used when evaluating the competing methods.

Stability of variable (gene) selection evaluated using 200 bootstrap samples. "# Genes": number of genes selected on the original data set. "# Genes boot.": median (1st quartile, 3rd quartile) of number of genes selected from on the bootstrap samples....

Effects of parameters of random forest on prediction error rate

Before examining gene selection, we first evaluated the effect of changes in parameters of random forest on its classification performance. Random forest returns a measure of error rate based on the out-of-bag cases for each fitted tree, the OOB error, and this is the measure of error we will use here to assess the effects of parameters. We examined whether the OOB error rate is substantially affected by changes in*mtry, ntree,* and *nodesize.*

For both real and simulated data, the relation of OOB error rate with *mtry* is largely independent of *ntree* (for *ntree* between 1000 and 40000) and *nodesize* (nodesizes 1 and 5). In addition, the default setting of *mtry* (*mtryFactor* = 1 in the figures) is often a good choice in terms of OOB error rate. In some cases, increasing *mtry* can lead to small decreases in error rate, and decreases in *mtry* often lead to increases in the error rate. This is specially the case with simulated data with very few relevant genes (with very few relevant genes, small *mtry* results in many trees being built that do not incorporate any of the relevant genes). Since the OOB error and the relation between OOB error and *mtry* do not change whether we use *nodesize* of 1 or 5, and because the increase in computing speed from using *nodesize* of 5 is inconsequential, all further analyses will use only the default *nodesize* = 1. These results show the robustness of random forest to changes in its parameters; nevertheless, to re-examine robustness of gene selection to these parameters, in the rest of the paper we will report results for different settings of *ntree* and *mtry* (and these results will again show the robustness of the gene selection results to changes in *ntree* and *mtry*).

Gene selection using random forest

Random forest returns several measures of variable importance. The most reliable measure is based on the decrease of classification accuracy when values of a variable in a node of a tree are permuted randomly, and this is the measure of variable importance that we will use in the rest of the paper. (In the Supplementary material we show that this measure of variable importance is not the same as a non-parametric statistic of difference between groups, such as could be obtained with a Kruskal-Wallis test). Other measures of variable importance are available, however, and future research should compare the performance of different measures of importance.

To select genes we iteratively fit random forests, at each iteration building a new forest after discarding those variables (genes) with the smallest variable importances; the selected set of genes is the one that yields the smallest OOB error rate. Note that in this section we are using OOB error to choose the final set of genes, not to obtain unbiased estimates of the error rate of this rule. Because of the iterative approach, the OOB error is biased down and cannot be used to asses the overall error rate of the approach, for reasons analogous to those leading to "selection bias". To assess prediction error rates we will use the bootstrap, not OOB error. In our algorithm we examine all forests that result from eliminating, iteratively, a fraction,*fraction.dropped*, of the genes (the least important ones) used in the previous iteration. By default,*fraction.dropped* = 0.2 which allows for relatively fast operation, is coherent with the idea of an "aggressive variable selection" approach, and increases the resolution as the number of genes considered becomes smaller. We do not recalculate variable importances at each step as mention severe overfitting resulting from recalculating variable importances. After fitting all forests, we examine the OOB error rates from all the fitted random forests. We choose the solution with the smallest number of genes whose error rate is within u standard errors of the minimum error rate of all forests. Setting $u = 0$ is the same as selecting the set of genes that leads to the smallest error rate. Setting $u = 1$ is similar to the common "1 s.e. rule", used in the classification trees literature [14,15]; this strategy can lead to solutions with fewer genes than selecting the solution with the smallest error rate, while achieving an error rate that is not different, within sampling error, from the "best solution". In this paper we will examine both the "1 s.e. rule" and the "0 s.e. rule".

On the simulated data sets backwards elimination often leads to very small sets of genes, often much smaller than the set of "true genes". The error rate of the variable selection procedure, estimated using the .632+ bootstrap method, indicates that the variable selection procedure does not lead to overfitting, and can achieve the objective of aggressively reducing the set of selected genes. In contrast, when the simplification procedure is applied to simulated data sets without signal (see Tables Tables11 and and22 in Additional file 1), the number of genes selected is consistently much larger and, as should be the case, the estimated error rate using the bootstrap corresponds to that achieved by always betting on the most probable class.

For additional results using different combinations of *ntree* = {2000, 5000, 20000}, *mtryFactor* = {1, 13}, *se*= {0, 1}, *fraction.dropped* = {0.2, 0.5}). Error rates when performing variable selection are in most cases comparable (within sampling error) to those from random forest without variable selection, and comparable also to the error rates from competing state-of-the-art prediction methods. The number of genes selected varies by data set, but generally the variable selection procedure leads to small (< 50) sets of predictor genes, often much smaller than those from competing approaches. There are no relevant differences in error rate related to differences in *mtry*, *ntree* or whether we use the "s.e. 1" or "s.e. 0" rules. The use of the "s.e. 1" rule, however, tends to result in smaller sets of selected genes.

Stability (uniqueness) of results

We have evaluated the stability of the variable selection procedure using the bootstrap. This allows us to asses how often a given gene, selected when running the variable selection procedure in the original sample, is selected when running the procedure on bootstrap samples. The results here will focus on the real microarray data sets. For other combinations of*ntree, mtryFactor, fraction.dropped, se*) shows the variation in the number of genes selected in bootstrap samples, and the frequency with which the genes selected in the original sample appear among the genes selected from the bootstrap samples. In most cases, there is a wide range in the number of genes selected; more importantly, the genes selected in the original samples are rarely selected in more than 50% of the bootstrap samples. These results are not strongly affected by variations in *ntree* or *mtry*; using the "s.e. 1" rule can lead, in some cases, to increased stability of the results.

Discussion

We have first presented an exhaustive evaluation of the performance of random forest for classification problems with microarray data, and shown it to be competitive with alternative methods, without requiring any fine-tuning of parameters or pre-selection of variables. The performance of random forest without variable selection is also equivalent to that of alternative approaches that fine-tune the variable selection process. We have then examined the performance of an approach for gene selection using random forest, and compared it to alternative approaches. Our results, using both simulated and real microarray data sets, show that this method of gene selection accomplishes the proposed objectives. Our method returns very small sets of genes compared to alternative variable selection methods, while retaining predictive performance. Our method of gene selection will not return sets of genes that are highly correlated, because they are redundant. This method will be most useful under two scenarios: a) when considering the design of diagnostic tools, where having a small set of probes is often desirable; b) to help understand the results from other gene selection approaches that return many genes, so as to understand which ones of those genes have the largest signal to noise ratio and could be used as surrogates for complex processes involving many correlated genes. A backwards elimination method, precursor to the one used here, has been already used to predict breast tumor type based on chromosomic alterations.

We have also thoroughly examined the effects of changes in the parameters of random forest (specifically *mtry, ntree, nodesize*) and the variable selection algorithm (*se, fraction.dropped*). Changes in these parameters have in most cases negligible effects, suggesting that the default values are often good options, but we can make some general recommendations. Time of execution of the code increases H" linearly with *ntree*. Larger *ntree* values lead to slightly more stable values of variable importances, but for the data sets examined, *ntree* = 2000 or *ntree* = 5000 seem quite adequate, with further increases having negligible effects. The change in *nodesize* from 1 to 5 has negligible effects, and thus its default setting of 1 is appropriate. For the backwards elimination algorithm, the parameter *fraction.dropped* can be adjusted to modify the resolution of the number of variable selected; smaller values of*fraction.dropped* lead to finer resolution in the examination of number of genes, but to slower execution of the code. Finally, the parameter *se* has also minor effects on the results of the backwards variable selection algorithm but a value of *se* = 1 leads to slightly more stable results and smaller sets of selected genes.

In contrast to other procedures our procedure does not require to pre-specify the number of genes to be used, but rather adaptively chooses the number of genes have conducted an evaluation of several gene selection algorithms, including genetic algorithms and various ranking methods; these authors show results for the Leukemia and NCI60 data sets, but the Leukemia results are not directly comparable since focus on a three-class problem. They report the best results with the NCI60 data set estimated with the .632 bootstrap rule. These best error rates are 0.408 for their evolutionary algorithm with 30 genes and 0.318 for 40 top-ranked genes. Using a number of genes slightly larger than us, these error rates are similar to ours; however, these are the best error rates achieved over a range of ranking methods and error rates, and not the result of a complete procedure that automatically determines the best number of genes and ranking scheme. A comparative study of feature selection and multi-class classification. Although they use four-fold cross-validation instead of the bootstrap to assess error rates, their results for three data sets common to both studies (Srbct, Lymphoma, NCI60) are similar to, or worse than, ours. In contrast to our method, their approach pre-selects a set of 150 genes for prediction and their best error rates are those over a set of seven different algorithms and eight different rank selection methods, where no algorithm or gene selection was consistently the best. In contrast, our results with one single algorithm and gene selection method (random forest) match or outperform their results.

Recently, several approaches that adaptively select the best number of genes or features have been reported. For the Leukemia data set our method consistently returns sets of two genes, similar to using an exhaustive search method, and lower than the numbers given by of 3 to 25 have proposed a Bayesian model averaging (BMA) approach for gene selection; comparing the results for the two common data sets between our study and theirs, in one case (Leukemia) our procedure returns a much smaller set of genes (2 vs. 15), whereas in another (Breast, 2 class) their BMA procedure returns 8 fewer genes (14 vs. 6); in contrast to BMA, however, our procedure does not require setting a limit in the maximum number of relevant genes to be selected have developed a method for gene selection and classification, LS Bound, related to least-squares SVMs; their method uses an initial pre-filtering (they choose 1000 initial genes) and is not clear how it could be applied to multi-class problems. The performance of their procedure with the leukemia data set is better than that reported by our method, but they use a total of 72 samples (the original 38 training plus the 34 validation of [44]) thus making these results hard to compare. With the colon data sets, however, their best performing results are not better than ours with a number of features that is similar to ours. Proposed two Bayesian classification algorithms that incorporate gene selection (though it is not clear how their algorithms can be used in multi-class problems).

The results for the Leukemia data set are not comparable to ours (as they use the validation set of 34 samples), but their results for the colon data set show error rates of 0.167 to 0.242, slightly larger than ours (although these authors used random partitions with 50 training and 12 testing samples instead of the .632+ bootstrap to assess error rate), with between 8 and 15 features selected (somewhat larger than those from random forest). Finally, [31], applied both shrunken centroids and a genetic algorithm + KNN technique to the NCI60 and Srcbt data sets; their results with shrunken centroids are similar to ours with that technique, but the genetic algorithm + KNN technique used larger sets of genes (155 and 72 for the NCI60 and Srbct, respectively) than variable selection with random forest using the suggested parameters. In summary, then, our proposed procedure matches or outperforms alternative approaches for gene selection in terms of error rate and number of genes selected, without any need to fine-tune parameters or preselect genes; in addition, this method is equally applicable to two-class and multi-class problems, and has software readily available. Thus, the newly proposed method is an ideal candidate for gene selection in classification problems with microarray data.

A reviewer has alerted us to the paper by Jiang et al., previously unknown to us. In fact, our approach is virtually the same as the one used by Jiang et al., with the exception that these authors recompute variable importances at each step (we do not do this in this paper, although the option is available in our code) and, more importantly, that their gene selection is based both in the OOB error, as well as the prediction error when the forest trained with one data set is applied to a second, independent, data set; thus, this approach for gene selection is not feasible when we only have one data set. Jiang et al. [45] also show the excellent performance of variable selection using random forest when applied to their data sets. The final issue addressed in this paper is instability or multiplicity of the selected sets of genes. From this point of view, the results are slightly disappointing. But so are the results of the competing methods. And so are the results of most examined methods so far with microarray data, as shown in [29] and [30] and discussed thoroughly by [27] for classification and by [28] for the related problem of the effect of threshold choice in gene selection. However, and except for the above cited papers and [6,46] and [5], this is an issue that still seems largely ignored in the microarray literature. As these papers and the statistical literature on variable selection (e.g., [40,47]) discusses, the causes of the problem are small sample sizes and the extremely small ratio of samples to variables (i.e., number of arrays to number of genes). Thus, we might need to learn to live with the problem, and try to assess the stability and robustness of our results by using a variety of gene selection features, and examining whether there is a subset of features that tends to be repeatedly selected. This concern is explicitly taken into account in our results, and facilities for examining this problem are part of our R code.

The multiplicity problem, however, does not need to result in large prediction errors. That the very different classifiers often lead to comparable and successful error rates with a variety of microarray data sets. Thus, although improving prediction rates is important, when trying to address questions of biological mechanism or discover therapeutic targets, probably a more challenging and relevant issue is to identify sets of genes with biological relevance. Two areas of future research are using random forest for the selection of potentially large sets of genes that include correlated genes, and improving the computational efficiency of these approaches; in the present work, we have used parallelization of the "embarrassingly parallelizable" tasks using MPI with the Rmpi and Snow packages [50,51] for R. In a broader context, further work is warranted on the stability properties and biological relevance of this and other gene-selection approaches, because the multiplicity problem casts doubts on the biological interpretability of most results based on a single run of one gene-selection approach.

EMERGING MODEL SYSTEMS IN PLANT BIOLOGY

The economic and ecological importance of forest trees provide the impetus for developing model systems to study tree biology. Wood is one of the most valuable commodities in the industrialized world, exceeding the weight of all other structural materials combined. More than half of the world's annual wood harvest is used as fuel, primarily in less-developed countries. In nature, forest trees are the principal form of terrestrial biomass. Wild forests provide irreplaceable environmental benefits such as carbon sequestration, watershed protection, and habitat for endangered wildlife. Trees themselves are intrinsically interesting, since they represent the pinnacle of evolutionary refinement in the competition among plants for access to sunlight, thereby dominating many of the Earth's most awe-inspiring landscapes.

While many aspects of tree biology are common to all plants, and hence can be studied in very tractable model species such as *Arabidopsis*, some unique facets of tree anatomy and physiology must be investigated in trees themselves. A large number of these unique characters are related to their perennial growth habit. Extensive formation of secondary xylem (wood) is perhaps the most obvious, but other characters include leaf and flower phenology, seasonal reallocation of nutrients, cold hardiness, iterative development of a complex crown form, and juvenile-mature phase change. The great lifespan and large size of forest trees pose special challenges to biologists, but are fundamental to understanding the role of trees in natural and managed ecosystems.

The experimental power of genetics, including molecular biology and genomics, has now permeated every discipline in biology. Clearly, an ideal model forest tree must have the potential to be manipulated using standard genetic approaches, such as mutagenesis and the creation of transgenic plants. In spite of the fact that other forest trees have been put forward as model systems, including *Salix* (willow), *Eucalyptus*, and *Pinus* (pine), for the reasons discussed below *Populus* has led the way in making use of genetic methods to study tree structure and function.

The knowledge gained from model forest tree systems will be put to use in the protection of forest ecosystems and in practical tree breeding. By defining, quantifying, and understanding tree physiological processes and morphological structures we hope to mathematically model, predict, and get a conceptual grasp of tree growth and development. In contrast to agronomic crops, relatively few physiologically based growth models have been developed for trees, largely due to a lack of fundamental physiological information and the difficulty of working with complex perennial plants (Ford 1985; Amaro and Tomé 1999). Just as in forest genetics, poplars have been adopted by tree physiologists as a model system.

The interplay between poplar genetics and physiology, at scales as large as the landscape and as small as a molecule, is the subject we now explore in greater depth. Ultimately, we wish to trace tree physiology and anatomy to the level of individual genes, understanding and manipulating their interactions with each other and the environment.

POPULUS TAXONOMY, DISTRIBUTION, AND GENERAL CHARACTERISTICS

Poplars are dioecious, wind-pollinated, and produce large amounts of pollen and small, cotton-tufted seed that is dispersed by wind and water in early summer. Capable of rapidly invading disturbed sites, many species occupy habitats in the dynamic environment of riverine floodplains where they form a key component of riparian forests (Braatne *et al.* 1996). Others, such as the aspens, commonly colonize upland areas after intense, stand-replacing fires (Burns and Honkala 1990). All poplars also have the capacity to reproduce asexually, mostly by sprouting from the root collar of killed trees, or from abscised or broken branches that become embedded in the soil. Aspen and white poplars propagate through sucker shoots that arise from horizontal roots, often after a fire, typically leading to clonal stands that may cover several hectares. Recurrent fires can maintain genets of such clones for hundreds of years.

Strengths of Populus as a Model System

Well established collaboration among poplar biologists. Extensive collaboration between poplar geneticists, physiologists, and pathologists has set a solid scientific foundation for joint efforts in the future. Shared genetic materials, genetically informative two- and three-generation pedigrees, DNA-based genetic markers, common field measurement protocols, clonal plantation trials supported by industry/government/academic partnerships, and funding from multi-agency grants have established a poplar research network of proven productivity. Working groups under the aegis of the International Poplar Commission (IPC) of the Food and Agriculture Organization (FAO) of the United Nations, and of the International Union of Forestry Research Organizations (IUFRO), as well as several university/industrial research cooperatives, help to coordinate the work and the exchange of information. Two books synthesizing a large body of data on poplar biology have been published recently, reviewing and solidifying the status of poplar as a model forest tree (Klopfenstein *et al.* 1997; Stettler *et al.* 1996).

Abundant genetic variation in natural populations. Adaptations to the diverse conditions inherent in large distribution ranges, as well as prominent genetic polymorphisms in local populations, offer poplar researchers a rich source of variation in tree morphology, anatomy, physiology, phenology, and response to biotic and abiotic stress. Much of this variation is under moderate to strong genetic control (Farmer 1996). Extensive natural populations remain for many species, especially in North America.

Ease of sexual propagation and interspecific hybridization. Poplars are bred in the greenhouse on detached female branches with pollen that can be stored for several years. Each pollination can yield hundreds of seeds within 4 to 8 weeks. Seeds germinate within 24 hours and give rise to 1-2 m tall seedlings by the end of the same year. Few if any trees can match such efficiency. Poplar species within the same section, and many of the species from different sections, can be hybridized. Because all members of the genus are diploid (2n = 38), hybrids are fertile and can generate F2 and backcross progenies that segregate for a wide range of traits. F1 hybrids often show heterosis in growth and associated characteristics which make them attractive for commercial use (Zsuffa *et al.* 1996).

Physiological responses to environmental variables are rapid and pronounced. As a consequence of the rapid juvenile growth of poplars, it is possible to measure short-term responses to biotic (*e.g.*, disease) and abiotic (*e.g.*, drought, elevated CO_2, ozone) factors. Opportunities are being explored for using this rapid growth as a means of carbon sequestration, particularly in Europe. The short turnaround time for physiological studies makes it possible to use molecular genetic approaches requiring very large sample sizes (hundreds or thousands of physiological measurements). High-throughput poplar physiology has led to the genetic mapping of quantitative trait loci (QTLs) controlling morphological (Bradshaw and Stettler 1995), phenological (Frewen *et al.* 2000), and pathological (Cervera *et al.* 1996; Newcombe and Bradshaw 1996; Newcombe *et al.* 1996; Villar *et al.* 1996) traits. Candidate genes for some of these QTLs have been identified (Frewen *et al.* 2000). Close coupling of physiological traits and biomass productivity. Critical components of productivity in *Populus* have been examined from a physiological perspective (Ceulemans 1990). Both structural and functional components were identified at different organizational levels: the individual leaf, the branch, and the whole tree. Among the process-related components, stomatal morphology and behavior (Tschaplinski *et al.* 1994), leaf morphology and leaf growth physiology (Ridge *et al.* 1986), leaf and whole tree photosynthesis (Isebrands *et al.* 1988), and root development (Friend *et al.* 1991) showed the most significant genetic variation. Key anatomical characters include those that determine whole tree leaf area and its duration, such as leaf demography (Chen *et al.* 1994; Hinckley *et al.* 1989), leaf size and distribution (Dunlap *et al.* 1992), branch size and distribution (Ceulemans *et al.* 1990), and leaf and branch orientation (Isebrands and Michael 1986). A favorable combination of several leaf physiological as well as whole tree and canopy structural traits explains the superior growth of selected hybrid poplar clones. QTLs for many of these traits have been mapped genetically (Bradshaw and Stettler 1995; Wu *et al.* 1997), but the genes controlling these traits remain to be identified and characterized.

Well-characterized molecular physiology. The large physical size of poplars, coupled with the well-understood movement of carbon assimilates (Vogelmann *et al.* 1982) and co-transported systemic wound signals between source and sink leaves (Davis *et al.* 1991; Clarke *et al.* 1998; Constabel and Ryan 1998) is an advantage over small plant models such as *Arabidopsis* in studies involving systemic signal movement. Induced resistance to insect herbivory is acquired systemically, and is phenocopied in part by mechanical injury (Havill and Raffa 1999). This positions poplar as an excellent model for synthesizing ecological and molecular perspectives of induced resistance to insect herbivory. Dramatic patterns of nitrogen allocation, utilization, and storage. Riparian ecosystems receive nutrient inputs episodically, which may help explain why poplars tissues store high levels of nitrogen in the form of vegetative storage proteins for subsequent utilization (Coleman *et al.* 1994; Lawrence *et al.* 1997). We have observed healthy poplar leaves with nitrogen concentrations exceeding 8% of dry weight (J. Cooke, K. Brown, J. Davis, unpublished), *vs.* the 1-1.5% maximum levels found even in fertilized conifer needles. Nitrogen levels fluctuate dynamically among organs within the same tree (roots, stem, and leaves) in concordance with seasonal rhythms of active growth and dormancy. The stem anatomy of poplar trees is particularly suited to studying the role of phloem-transmissible substances such as glutamine in regulating nitrogen allocation, as phloem can be specifically perturbed by girdling while xylem transport remains intact (*e.g.*, Sauter and Neumann 1994).

Physiological process models for poplar growth and development. Large databases of anatomical, physiological, and silvicultural traits are available for a modest number of *Populus* hybrids and clones. Several physiologically-based growth and productivity models have been developed from these data. The models include basic information on carbon uptake and allocation in poplar, as well as components and parameters of leaf display and crown structure. Most of the models simulate carbon uptake, carbon allocation, growth, and/or light interception in poplar and incorporate some specific parameters of leaf display, position in the tree, and branch structure (Chen *et al.* 1994; Host *et al.* 1996; Isebrands *et al.* 1996). Data on the physiological and structural growth determinants at the leaf, branch, and whole tree level indicate that differences in clonal productivity can be incorporated into the ideotype concept developed for poplar tree breeding under short rotation intensive culture (Dickmann 1985; Dickmann and Keathley 1996).

Cloning of individual tree genotypes. The ease with which most materials can be vegetatively propagated is one of poplar's premier assets. Cloning captures genetic variation and allows it to be replicated in space and time in separate experiments. Cloning 'freezes' genetic variation in hybrids and permits the side-by-side growth of multiple generations of a pedigree. Cloning permits the growth of abnormal plants under field conditions that in the competitive environment of a seedling population would be impossible. Cloning also allows destructive sampling for physiological studies, the sharing of materials among laboratories, and the buildup of cumulative knowledge on selected genotypes.

Closely related to other angiosperm model plants. Unlike the pines and other gymnosperms, poplars are relatively recently diverged from other angiosperms, such as *Arabidopsis,* which serve as models for integrating genetics into the study of plant biology (Fig. 1).

Small genome size. The haploid genome size of *Populus* is 550 million base pairs (bp) (Bradshaw and Stettler 1993), only 4 times larger than the genome of the model plant *Arabidopsis,* and 40 times smaller than the genomes of conifers such as loblolly pine. The small poplar genome simplifies gene cloning, Southern blotting, and other standard molecular genetics techniques. The physical:genetic distance ratio for poplar is close to 200 kb/centiMorgan, almost identical to *Arabidopsis,* making *Populus* an attractive target for map-based (positional) cloning of genes.

Basic molecular genetics toolkit. Linkage maps of the poplar genome have been made with a variety of marker types, including allozymes, RFLPs, STS/CAPs, RAPDs, AFLPs, and microsatellites (Liu and Furnier 1993; Bradshaw *et al.* 1994; Cervera *et al.* 1996; Frewen *et al.* 2000). Bacterial artificial chromosome (BAC) genomic libraries containing 10 haploid poplar genome equivalents (50,000 clones) have been constructed for *P. deltoides* (W. Boerjan, pers. comm.) and for *P. trichocarpa* (J. Vrebalov, pers. comm.). These libraries have been arrayed into microtiter plates and replicated for sharing with other researchers. Expressed sequence tag (EST) databases of modest size are in the public domain (Sterky *et al.* 1998). Facile transformation and regeneration to create transgenic poplars. The ease of creating transgenic *Populus* is unmatched by any other forest tree. Some poplars, such as the hybrid aspen (*P. alba* x *tremula*) clone 717-1B4, can be genetically transformed with *Agrobacterium* and regenerated efficiently into intact transgenic trees within 6-10 months (Jouanin *et al.* 1993). Transgenesis is the 'gold standard' for demonstration of gene function, and so is crucial for basic research into physiological processes.

Obviously, there are also potential commercial applications for transgenic poplars. The ease with which genes are cloned and transgenic trees are produced in *Populus,* set against the backdrop of a poplar EST database of wood-forming genes (Sterky *et al.* 1998), has created a fertile scientific field for understanding and manipulating wood composition in dramatic ways (Hu *et al.* 1999). Progress is likely to accelerate in this important research area as the tools of genomics, biochemistry, and cell biology are collectively brought to bear on manipulating wood quality and other important physiological or morphological traits.

Weaknesses of Populus as a Model System

Modest commercial importance. Poplar is grown commercially on a wide geographic scale and many wild populations, especially aspen in the boreal forest, are harvested by industry. However, from a corporate viewpoint, poplars are much less important than pines and other softwoods, or eucalypts. It can be difficult to persuade companies to invest in (or lobby for) poplar research, either as a model tree, or as a component of fiber/energy plantations. For a variety of historical, political, and economic reasons, most of the forest products industry still relies on harvests from natural or extensively managed stands and has yet to turn to the intensive agricultural paradigm for wood production for which poplar is ideally suited.

Long generation interval. Many willows and some eucalypts can be induced to flower within a year or two from seed. Most poplars will not flower earlier than 4 years of age, and many will take twice that long. The long generation interval is an impediment to practical breeding and selection and the development of informative pedigrees. It limits the applicability of conventional experimental genetic techniques such as induced mutagenesis, since producing homozygotes from heterozygous mutants is impracticably slow. Outcrossing mating system. Poplars are dioecious, so self-pollinations cannot be done (with a very few hermaphroditic exceptions). Classical genetic tools such as inbred lines cannot be produced rapidly enough to be useful. Large physical size of trees. There is no such thing as a 'small experiment' when one works with poplars! Beyond one year of age the physiologically active crown is above human reach, requiring towers or remote sensing. Lack of well-planned breeding programs. The advantages of cloning have been pointed out, but there is a hidden disadvantage. Most poplar breeding programs have relied upon more-or-less random interspecific hybridization to produce large progeny arrays, followed by very intense clonal selection to identify the best clones out of perhaps 10,000 seedlings. While demonstrably effective at producing remarkable genetic gains in short order, the scarcity of well-planned, long-term, sustained breeding programs and associated breeding materials hampers genetic studies. For example, multi-generation pedigrees are the basis for genetic mapping experiments designed to identify QTLs affecting important tree phenotypes, yet few such pedigrees exist.

What is needed to increase the value of populus as a model system?

Long-term support for germplasm collection and maintenance. Nothing is more destructive to sustained progress on long-lived plants than inconsistently funding basic genetic infrastructure such as breeding, maintenance of living clone banks, and annual measurement of experimental plantations. Unfortunately, these routine but important activities are difficult to justify to most funding agencies and usually are jettisoned at the first dip in finances. A solution to this problem might be to establish at least one public genetic stock center on each continent with an active interest in poplar research. This would promote the centuries-old practice of exchanging pollen, seed, and cuttings from genetic stocks with value for research and commercial use. The poplar research community's tradition of accumulating and integrating knowledge about shared pedigrees would be encouraged.

Shorter generation interval; early or inducible flowering. Delayed flowering represents a lost opportunity for tree geneticists to provide suitable pedigrees for physiological studies, and is currently the single most limiting factor in the conventional genetic improvement of forest trees. The development of methods for inducible early flowering, either by chemical/ physical treatment or genetic engineering, will make it possible to decrease the generation interval for purposes of breeding, while also preventing excessive flowering from siphoning photosynthate away from wood formation in production plantations. While some progress has been made in this area with poplars (Weigel and Nilsson 1995), much remains to be done (Rottman *et al*. 2000). Complete genomics toolkit. Probably no other group of organisms will benefit more from the development of genomics tools than forest trees, where conventional genetic approaches are so limited. Since most of the interesting traits in trees are inherited quantitatively, a dense genetic map composed of 2500-5000 genetic markers which identify homologous loci in all *Populus* species and have a high information content, such as microsatellites or single-nucleotide polymorphisms (SNPs), should be produced specifically for QTL mapping.

The genome of *Populus* is small enough to invite the construction of a complete physical map of overlapping BACs. The physical map could be tied to the genetic map either by mapping genetic markers onto the physical map, or by developing genetic markers directly from BAC sequences having known physical positions. The physical and genetic maps would serve several important purposes: as a starting point for map-based gene cloning, as a means to determine the extent of collinearity between the *Populus* genome and that of the completely sequenced genome of *Arabidopsis* (thus making effective use of the growing knowledge of *Arabidopsis* genome structure and function), and as a scaffold for the eventual sequencing of the entire *Populus* genome.

Until the *Populus* genome is completely sequenced, perhaps in the next 5-10 years, a comprehensive EST database will provide a source of SNPs, and become the foundation for microarray development and use for studying global patterns of gene expression. EST microarrays will allow poplar researchers to couple physiological measurements with gene expression profiles, illuminating the genes, biochemical pathways, and cellular processes that are affected by a given environmental perturbation or transgene.

Steve Strauss (pers. comm.) has suggested an ambitious strategy of discovering genes with novel functions in poplar by generating 50,000 dominant gain-of-function mutants in transgenic poplars by activation tagging (Walden *et al*. 1994), which is an insertional mutagenesis strategy that is more suitable for trees with their outcrossing mating systems than traditional mutagenesis approaches that yield mostly recessive mutants. The incorporation of such tools will be important if we are to carry out forward and reverse genetics with equal facility in forest trees.

GENETIC APPROACH OF STEM SECONDARY GROWTH

Wood is the largest fraction the biomass on this planet. In addition to the primary meristems for development of primary tissues, tree species also produce secondary tissue (wood) from the vascular cambium in a process called secondary growth, consisting of successive addition of secondary xylem which is responsible for radial expansion of woody stems. Several zones can be observed according to function in a stem cross section. The pith is central with adjacent xylem (wood), composed mainly of dead cells involved in the transit of water and minerals[5]. Bark makes up the outer zone, with adjacent phloem, where sap is transported. Between the xylem and the phloem is the vascular cambium. Thus from the center to the periphery is the pith, wood (xylem), cambium, phloem and bark. The cambial meristem produces both new xylem and phloem in a bi-directional manner, which in time form the wood and bark, respectively. Differentiation of xylem cells from procambial cells is coupled with expression of Class III HD-ZIP genes. In *Arabidopsis*, this gene family contains *AtHB8* and *AtHB15* that are expressed in vascular bundles, in particular, in the procambial or xylem precursor region.

Vascular bundles present in many plant leaves show a distinct dorsoventral organization - xylem is located on the dorsal (adaxial) side, phloem is on the ventral (abaxial) side with procambium positioned between xylem and phloem. *Antirrhinum majus* and *A. thaliana* mutants that show defects in the identity or maintenance of the dorsoventral axis are indicative of the involvement of the dorsoventral axis in the radial-pattern formation of leaf vascular tissues. There is a relationship between vascular and overall body patterning as mutations disrupting directional growth such as establishment of new growth axes, usually, also interfere with vascular strand formation. For example, the *Arabidopsis* triple mutant of the Class III HD-ZIP genes; *rev phb phv,* develops radially symmetrical cotyledons with abaxialized vasculature, in which phloem develops on the adaxial and xylem on the adaxial side. In contrast to the normal collateral pattern where phloem is on the abaxial side and xylem is on the adaxial side.

Within the vascular strands, the xylem and phloem tissues differentiate asymmetrically from the procambium. The patterning of the vascular bundles in the shoot is closely associated with adaxial/abaxial patterning of the lateral organs and with the establishment of central versus peripheral identity within the stem. Dominant *rev-d, phb-d* or *phv-d* gain-of-function mutants have adaxialized lateral organs and develop vascular bundles in which xylem surrounds phloem. All of the described Class-III HD-ZIP gain-of-function mutations are located in a microRNA165 (miR165) or miR166 target sequence within a putative sterol /lipid-biding START domain sequence. Mutations that disrupt the miRNA in a similar gain-of-function phenotype, suggests that the *HD-ZIP-III* genes are under the control miRNA regulation.

YABBY gene family is another gene family responsible for adaxial/abaxial patterning in lateral organs. It expression is located in the abaxial side of the leaves. *KANADI* gene family members also contribute to abaxial identity in leaves, and are also active in the stem, where YABBY gene family members are not active. Triple loss-of-function mutants of *kan1 kan2 kan3* show radialized amphivasal vascular bundles in stems. The *kan1 kan2 kan3* mutants, phenocopies the *rev-d, phb-d* or *phv-d* gain-of-function mutants, and has abaxial Class III HD-Zip gene expression.

While several plant hormones have been implicated in the regulation of vascular tissue formation, considerable evidence indicates that auxin is the major signal involved in several aspects of the ontogeny of the vascular system. When supported by cytokinins, indole-3-acetic acid (IAA) can induce xylem tracheary element differentiation in suspension culture cells of suitable species. The findings of Ko and colleges after microarrays analyses of secondary growth induced in *Arabidopsis* plants, suggests that many of the genes up-regulated in wood-forming stems might be regulated by auxin. It is clear that polar auxin transport is crucial for the continuous formation of vascular bundles, however, the precise molecular mechanisms are still mostly unknown. To understand better these processes, we need to identify more components that are involved in auxin-flow-dependent procambial cell differentiation. Brassinosteroids (BRs) also influence the vascular system. It has been shown that endogenous biosynthesized BRs promote xylem formation and suppress phloem formation. Late studies have shown that *HD-Zip-III* genes seem to be regulated by BRs in vascular cells. The vascular differentiation defects seen in BR-deficient mutants form increased amounts of phloem and reduced amounts of xylem. Mutant analysis of the three BR receptors suggests that they act redundantly to regulate vascular differentiation. Possibly signaling through the BR receptors in the procambium induces xylem proliferation and, at the same time, represses phloem proliferation.

Key genetic mechanisms originally characterized as regulating the meristematic cells of the shoot apical meristem are also expressed in vascular cambium during wood growth. Both KNOX and MYB transcription factors have been shown to regulate genes that encode key cell wall enzymes. Using xylem and phloem whole transcriptome studies from secondary tissues of *Arabidopsis,* Zhao et al. realized that the same genes involved in the regulation of primary meristems, are also regulators of the cambium. *CLAVATA1* and 2 were expressed in the phloem and cambium but not *CLAVATA* (*CLV3*). Although, genes belong to the CLE-family, from which *CLV3* belongs to, were expressed in this tissues.

5

Application of Plant Tissue Culture to Agriculture and Forestry

The most important challenge in a developing country is food for its growing population. The arable land area is fixed. Some means has to be there to increase the productivity. Plant tissue culture technique offers an excellent opportunity for mass propagation of plants in laboratory test tubes, which are transferred to the field. Besides crop plants, the technique is also applied to regenerate saplings for plantation and regeneration of dwindled forests. Some rare and nearly extinct plant species can be rescued and propagated by this technique.

Embryos produced by incompatible crosses also are rescued, seed dormancy is overcome, life cycle is shortened and much more. This technique is amalgamated with genetic engineering to regenerate plants with novel characters and combine two or more beneficial characters into a single plant. Things, once seemed impossible, have been made possible. We shall discuss about four elementary applications of this technique, namely, micro-propagation; organogenesis; somatic embryogenesis and protoplast culture and fusion.

MICRO PROPAGATION

Basically, micro propagation is similar to rooting of plant cuttings and is, in a way, another method of vegetative propagation of plants. However, it differs from the conventional procedure in that it is carried out in an aseptic condition and requires a unique recipe i.e. an artificial nutrient medium. It is used for forestry improvement and is an example of direct laboratory to land transfer of biotechnological benefits A small plant cutting or explant (usually an axillary bud) is surface sterilized and inoculated into a culture vessel containing a semi-solid nutrient medium. The inoculated culture vessel is incubated at room temperature.

In a day or two, a large number of shoots develop from the axillary bud in a process known as axillary bud proliferation. Each growing point is sub-cultured to give rise to shoot. This phenomenon is known as adventitious shoot formation. Each shoot is stimulated by an auxin to develop roots. The new plantlet is transferred to the field.

Organogenesis

Organogenesis, in essence, refers to differentiation of organs, such as shoot and root from an undifferentiated mass of cells. The cells of an explant are highly differentiated. When an explant is placed in an artificially enriched nutrient medium, its differentiated cells first de-differentiate and form a mass of unorganized cells known as callus. The cells of the callus then re-differentiate and produced the desired tissue and then an organ or organs under the influence of specific growth regulators (hormones). Single cells also can be cultured and made to develop shoot and root one followed by the other. Plant growth regulators (hormones) play an important role in this regard.

There are two important groups of plant hormones: cytokinins and auxins. Cytokinins such as kinetin and adenine promote shoot differentiation, while auxins, such as indole acetic acid (IAA) and naphthalene acetic acid (NAA) promote root differentiation. It has been established that shoo root differentiation depends upon the ratio or quantitative interaction betwee cytokinins and auxins. For example, two molecules of kinetin or 15,000 molecule of adenine are required to neutralize one molecule of IAA. Thus, shoot /10 differentiations is a function of the quantitative interaction between cytokinin an auxin. This principle is applied to plant cells or tissues cultured in vitro. Kinetin and IA are added to the in vitro culture in required amounts, one following the others- promote shoot and root differentiation.

Somatic embryogenesis

In flowering plants, an embryo is the product of the zygote and the zygote is the product of the fusion of two gametes. The embryo undergoes a preprogrammed development and forms a plantlet. In a nutshell, under a normal circumstance, a embryo is the result of sexual reproduction. Such embryos are known as zygoti embryos. However, plant tissue culture technique offers a method of producing; embryos from somatic cells bypassing sexual reproduction such embryos a known as somatic embryos. The process of formation of somatic embryos is known as somatic embryogenesis.

Somatic embryo formation starts with a mass of single cells or a tissue grown on a semirsolid nutrient medium. A cell repeatedly divides and forms a cell aggregate. The cell aggregate passes through different stages, such as globular, heart shaped and torpedo shaped stages. The torpedo shaped stage is the mature stage. The culture is initially started on a semi-solid medium and the callus so formed is transferred to a liquid medium in an agitated and aerated bioreactor. The cells breaking off from the callus develop into somatic embryos. Mature stages (torpedo shaped stages) are sorted out and grown to maturity on a semi-solid medium. Dormancy is induced before these are processed for transfer to the field. There are four methods, by which the somatic embryos are transferred to the field.

1. These are germinated in the laboratory, transplanted in pots and then transferred to the field.
2. The dormant embryos are encapsulated in a gel containing an adequate nutrient for the embryo. These encapsulated embryos are known as artificial or synthetic seeds such seeds can be planted in the field.
3. This method involves the germination of the embryos under a controlled condition and then mixing of the seedlings is mixed with a gel like medium. The seedling- gel mix is sown in the field.
4. The embryos are germinated and then fluid drilled.

The major advantages of this process are: (1) rescue of zygotic embryos formed by: incompatible crosses and (2) overcoming seed sterility and dormancy.

Protoplast culture and fusion

Protoplasts are plant cells, whose cell wall is digested. The cell is bound by a plasma membrane. For the isolation and culture of protoplasts, see section isolated protoplasts from two different species of plants have been successfully fused to produce a single protoplast containing the genetic material and the cytoplasm of both the fusing protoplasts. The process of fusion is not straightforward It is facilitated by some agents, known as fusogens. Fusogens are of two types: chemical and electrical. Poly ethylene glycol (PEG) is a chemical fusogen.

It is not used as a universal fusogen, since it is toxic to protoplasts of some plants; alternately, pulses of electric current (direct current) are applied to the fusing protoplasts. This method is known as electro-fusion. The depicts the sequential fusion of two protoplasts, resulting in a synkaryon. There are three steps in this process. First, two protoplasts come in close proximity. Then the plasma membranes of the two fuse and then the two nuclei lie in the mingled cytoplasm. This stage is known as a heterokaryon. In the third stage, the two nuclei fuse forming a synkaryon.

The cell wall of the fused protoplast is regenerated and the cell is cultured in an artificially enriched nutrient medium. The following sequence of events is same as that in the callus culture. This process is also known as somatic hybridization and the products as somatic hybrids. This method bypasses the fusion of gametes of two unrelated plant species. P. S. Carlson, H. H. Smith and R. D. Dearing (1972) obtained the first somatic hybrid by fusing the isolated protoplasts of Nicotiana glauca with

N. langsdorfii. However, sometimes, due to cellular incompatibility, two nuclei can not co-exist. Consequently, one nucleus is eliminated and a protoplast containing the nucleus of one species and the cytoplasm of both results.

The ensuing hybrid is known as a cytoplasmic hybrid or cybrid. Somatic hybridization is attempted in plant species, which are sexually incompatible. The well known example of a somatic hybrid is 'pomato' obtained from the protoplasts of potato and tomato. However, this hybrid is of little commercial value.

TISSUE CULTURE

Term biotechnology is the science of applied biological processes. The most widely accepted definition of biotechnology is the integrated use of biochemistry, microbiology and engineering sciences in order to achieve technology (industrial) application of the capabilities of micro-organisms, cultured tissue cells and parts thereof. Tissue culture is increasingly applied to wide range of biotochnology ventures especially to the micropropagation and genetic improvement of crop plants. The techniques of protoplast culture and fusion are of special significance for improvement of crops such as potato, sugarcane, cassava and banana which were vegetatively propagated. Somatic cell fusions have been successfully employed for the resynthesis of rapeseed (*Brassica napus*) by fusing protoplasts of *Brassica oleracea* and *B. campestris* and for developing hybrids between *Medicago sativa* and *M. julcata*. Tissue culture" technique has a lot of scope in improving the crop cultivars, forestry and animal sciences. In developing high yielding hybrid, resistant to insect pests and diseases, resistant to drought and to transfer any other desirable character into another organisms, the tissue culture technique playa major role. Having acquired greater depth in the knoledge of tissue cellular and genetic systems operating in the biosphere man has developed biotechnologies which can be used in medicine, industrial and in agriculture to enhance quality and quantity in .all spheres of production and processing. Todays biotechnology has a long history as ancient as any other biological sciences.

HISTORY

In the history of biotechnology several stages of development can be identified as given in table 1 below.

Table. Historical development of biotechnology Year Development

Sl.No	Year	Development
1	Before 6000 B.C	Yeast employed to produce bear and wine
2	Before 3000 B.C	Providing bread with leaven, fermentation of juices to alcoholic beverages
3	34 d Century B.C	Manufacture of beer in Sumeria, Babylonia and Egypt

4	From 1150 A.D	Manufacture of ethanol
5	14th century	Vinegar manufacturing industry New Orleans
6	After about 1650	Artificial cultivation of mushroom in France
7	Before 1670	Copper mined with the help of micro organisms
8	1677	Antony Van Leeuwenhoek first observed the yeast cells
9	1748	Observations on the generation composition and decomposition of animal and vegetable substances
10	1818	Discovery of the fermentation properties of yeasts by Erxleben
11	1857	Lois Pasteur described Lactic acid fermentation
12	1869	F . Miescher isolated nucleic acid
13	1878	Joseph Lister reported lactic acid fermentation and its bearing on pathology
14	1881	Production of lactic acid from micro organisms
15	1890	Alcohol first used to fuel motors
16	1890	S. Kitasato and E. Von Behring observed antibodies
17	1897	Eduard Buchner observed alcoholic fermentation without yeast cells
18	1910	Establishment of large scale sewage purification system
19	1915	German process for the manufacture of backery yeast
20	1920	Production of citric acid in a surface process
21	1928	Discovery of penicillin by Alexander Flemming
22	1944	Large scale production of penicillin discovery of Streptomycin
23	1946	Genetic recombination in bacteriophage
24	1953	Double stranded nature of DNA revealed
25	1962	Mining of Uranium with the help of microbes
26	1971	Transfer of nif gene from Rhizobium to Klebsiella
27	1973	First successful genetic engineering experiment done
28	1975	Discovery of hybridoma technique for production of monoclonal antibodies
29	1978	Restriction enzymes and their application to molecular biology
30	1981	US approval for using monoclonal antibodies in diagnosis
31	1982	Genetically engineered insulin approved for use
32	1984	Animal interferons approved for use against cattle disease
33	Mid Eighties	Use of interferons for the treatment of viral disease. Genetically engineered human growth hormone

		approved for treatment of dwarfism. Wide use of monoclonal antibodies in diagnosis. Introduction of genetically engineered hepatitis vaccine. Production of new antibodies
34	Late Eighties	Use of interferon for treatment of hairy cell leukemia. Raw materials obtained from microbes for plastic industry. Large scale production of industrial chemicals using microbes
35	Nineties	Extraction of oil from the ground with the help of genetically engineered microbes. Extraction of metals from factory waste. Wide use of recombinant DNA technology in medicine, human health, agriculture and industry

Biotechnology is not new science rather it is fast growing field of science. It includes branches like molecular biology, tissue culture, genetic engineering and plant pathology. Biotechnology is developing in close collaboration with these branches of science and can develop further only by means of an interdisciplinary approach between them. During the last two decads plant cell, tissue and organ culture have developed rapidly and become a major biotechnological tool in agriculture, horticulture, forestry and industry. Those problems which were not solved through the conventional techniques, are now have being solved by these techniques; for example inter- and intra-specific crosses, micro-propagation, somaclonal variation, encapsulated seeds etc. The theory behind tissue culture is that "each living cell, of a multicellular organisms, is capable of independent development, when provided with suitable conditions." This term was coined as "totipotency". The concept of totipotency is 'important in tissue culture. Use of multicellular organisms in research, as biological units, is rather difficult, therefore, attempts to study an organism by reducing to its constituent cells and subsequently the cultured cells as basic organisms are of fundamental importance.

In plant tissue cultural experiments, organs, plant tissue: and cells are isolated from the plant and made to grow in different types of culturing vessels, viz., flasks, bottles, tubes and watch glasses containing essential mineral salts, organic compounds such as amino acids and vitamins and auxin or growth substances e.g. Indole Acetic Acid (1AA),Indole Butyric Acid (IBA), 2,4-D, cytokinins and gibberellines in various concentrations and combinations, Since the green cells also become dependent on external supply of carbohydrates, the culture medium must also contain a readily metabolizable carbon sources, a sugar.

COMPOSITION AND MATERIAL

Tissue culture can be maintained either in liquid medium or semi-solid medium in the former, plant material is immersed in the medium either partially or completely while in the latter plant material is placed on the surface of the medium. The principal components of most plant tissue culture media are inorganic nutrients (macronutrients and micronutrients, carbon sources, organic supplements, growth relulators and a gelling agent

Inorganic nutrients

A number of inorganic nutrients are required for normal growth of plants. Besides carbon, hydrogen, oxygen, other elements essential for plant growth include nitrogen, Phosphorucs, potassium, calcium, sulphur, magnesium, iron, manganese, copper, boron, zinc and molybdenum. Iron is added in the form of Ferric sulphate while nitrogen is furnished in the form of nitrate and ammonia. Of all the mineral nutrients, nitrogen is responsible for the most pronounced effects on growth and differentiation of cultured tissues.

Organic nutrients

Cultured plants accomplish better growth when medium is supplemented with organic nutrients such as amino acids and vitamins. The most commonly used vitamin is thiamine (vitamin B_2). Other vitamins which improve the growth of cultured plants include nicotinic acid acid pantothenate and pyridoxine. Besides these boitin, folic acid and ammobenzoic acld are added in various concentrations in to some of the media. In order to promote growth of callus, some complex semi-synthetic substances are added in the medium. They include yeast extract (YE), coconut milk (CM), Casein Hydrolysate (CH), tomato juice (TJ) and malt extract (ME). As a carbon source, sucrose is the most preferred carbohydrate. Glucose, fructose, maltose, galactose, mannos and lactose are other favourable sugars.

Growth Hormones

Growth hormones helps in better plant development. The most commonly used auxins are IAA, IBA, NAA (naphthalene acetic acid) and 2,4-D (dichlorophenoxyacetic acid). These auxins promote cell division and root differentiation. Cytokinins are responsible for cell division and shoot differentiation. BAP (benzyl-amino purine), and 2-isopentanyl adenine are most widely used cytokinins. Among gibberellines, GA is commonly used.

Agar

Agar (a polysaccharide obtained from sea weeds) is used to solidify the medium. Agar is commonly used at a concentration of 0.8-1% (W/v). Use of higher concentration of agar makes the medium hard This prevents the diffusion of nutrients into the tissues.

pH

The pH of the medium is adjusted between 5 to 5.8 by adding (0.1 N) NaOH or HO. Usually a pH higher than six results in a fairly hard medium whereas a pH below five does not allow satisfactory solidification of the medium.

MEDIA PREPARATION

The simplest method of preparing culture media is to use commercially available powdered media. These media contain all the requisite nutrients. However, agar, sugar and other supplements are added in the finally prepared media.

For the preparation of medium, the following steps are followed.

1. Appropriate quantities of agar-agar and sucrose are dissolved in distilled water.
2. Required quantities of stock solutions, growth hormones and other supplements are added
3. More distilled water is added to make the final volume of the medium.
4. pH of the medium is adjusted between 5-5.8 by adding 0.1 N NaOH or 0.1 N HCl.
5. The medium is poured in culture tubes, flasks or any other containers.
6. Culture vessels are plugged with non-absorbent cotton wool wrapped in cheese clothes. This will allow free gaseous exchange but inhibits microbial contamination.
7. Culture vessels are sterilized by autoclaving at 120°C (1.06 kg/cm2) for 15-20 minutes.
8. Culture medium is allowed to cool at room temperature and used or stored at 4°C.

INOCULATION AND CULTURING

The process of inoculation consists of sterilization and isolation of single cells with which the medium is inoculated. This is followed by careful culturing the inoculum.

Sterilization

Depending upon the type of sterilization and material to be sterilized. There are four means of sterilization.

Dry heat

Glass wares (flasks, tubes, filters and pipettes) metal instruments and other articles which do not get charred by high temperature are put in special containers or wrapped in thick brown paper or thick aluminium foil and placed in drying oven and sterilized for a period of not less than 4 hours at a temperature of 140-160°C and then taken to the transfer chamber.

Wet heat

In this process an autoclave operated with water vapour under pressure (steam pressure of 151b/inch2) and at a temperature of 120°C applied for 20 minutes when the autoclave has reached proper temperature mid the residual air enclosed inside the autoclave has been displaced by steam. "Wet" heat sterilization process is recommended for heat stable solutions. Rubber items or other partly heat labile articles must not be autoclaved.

Filtration

Some protein material, vitamins and growth substances are heat labile and cannot be autoclaved and, therefore, they must be sterilized by filtration using ultrafilter at room temperature, before being added to an agar medium. The solution of these substances are added to the autoclaved medium with a sterile pipette while the agar medium is cooling and still in the sol state. In using this procedure the filter pore size is given utmost importance. A pore-size of 0.22 mm has been recommended in Millipore Catalogue and pure chasing guide 1978/79. All the stopcocks of the filtration unit are opened before it is wrapped in thick brown paper and autoclaved. Media or its constituents can also be sterilized by passing them through a pyrex sintered -glass sterilizing filter (porosity H5).

Sterilization by chemicals

Surgical blades, scalpel are not sterilized by dry heat, because the high temperature makes the cutting edges dull. These articles as well as spatulas and forceps are usually immersed in 80% V/v, ethyl alcohol until required, and sterilized during use by frequent immersion in alcohol and flaming. Before the aseptic removal of plant organ or tissue explant it is necessary to surface sterilize the plant material. Surface sterilization may be accomplished by using an appropriate surface sterilant. Generally aqueous solutions of calcium or sodium hypochlorite ($Ca(OCl)_2$ or NaOCl), which release chlorine as the active sterilant are used. Some other chemicals which have also been used are H_2O_2, bromine water, silver nitrate and mercuric chloride.

Isolation of Single cells

Isolation of cells are done in several methods which are explained here.

From Plant Organs

The most suitable material for isolation of single cells is the leaf tissue since a more or less homogeneous population of cells are there in the leaves. There are good for raising defined and controlled large-scale cell cultures. From such intact plant organs (as leaf tissue) single cell can be isolated using mechanical or enzymatic method.

- *Mechanical method:* The procedure involves mild maceration of 10 g leaves in 40 m1 of the grinding medium (20 u mol sucrose, 10 u mol $MgCl_2$, 20u mol tris-HCI buffer with pH 7.8) with a mortar and pestle. The homogenate is passed through two layers of tmuslin cloth and the cells thus released are washed by centrifugation at low speed using the same medium. Isolation of free parenchymatous cells can also be achieved on a large scale by the mechanical method.
- *Enzymatic method:* Isolation of single cells by the enzymatic method has been found convenient as it is possible to obtain high yields from preparations of spongy parenchyma with minimum damage or injury to the cells. This can be accomplished by providing osmotic protection to the cells. while the enzyme macerozyme degrades the middle lamella and cell wall of the parenchymatous tissue. Applying the enyzmatic method to cereals has proven difficult since the mesophyll cells of these plants are apparently elongated with a number of interlock constrictions, thereby preventing their isolation.

From cultured tissues

The most widely applied approach is to obtain a single cell system from cultured tissues. Freshly cut pieces from surface-sterilized plant organs are simply placed on a nutrient medium (solidified) consisting of a suitable proportion of auxins and cytokinins to initiate cultures. Explants on such a medium exhibit callusing at the cut ends, which gradually extends to the entire surface of the tissue. The callus is separated from an explant and transferred to a fresh medium of the same composition to enable it to build up a mass of tissue. Repeated subculture on an agar medium improves the friability of the callus, a pre-requisite for raising a fine cell suspension in a liquid medium.

The pieces of undifferentiated and friable callus are transferred in a continuously agitated liquid medium dispensed in autoclaved flasks or other suitable vials. Agitation is done by placing the culture flasks/vials on an orbital platform shaker or any other suitable device. Movement of the culture medium exerts mild pressure on small pieces of tissue, breaking them into free cells and small cell aggregates. Further, it augments the gaseous exchange between the culture medium and the culture air, and also ensures uniform distribution of cells as well as cell clumps in the medium.

Growth and maintenance

Cell suspensions are clonally maintained by the routine transfer (subculture) of cells in the early stationary phase to a fresh medium. During the incubation period the biomass of the suspension cultures increases due to cell division and cell enlargement This continues for a limited period since the viability of cells in suspension after the stationary phase de- creases due to the exhaustion of some factors or the accumulation of toxic substances in the medium. At this stage an aliquot of the cell suspension with uniformly dispersed free cells and cell aggregates is transferred to a fresh liquid medium of the : original composition.

The timing of subcultures is very important. The incubation period from cultures initiation to the stationary phase is determined primarily by: (a) initial cell density, (b) duration of lag phase and (c) growth rate of cell line. The cell density used to subculture is critical and depends largely on the type of suspension culture to be maintained. Low initial cell densities will prolong the lag phase and exponential phases of growth. While initiating new suspension culture it is necessary to determine optimal cell density, proportionate to the volume of the culture medium, in order to achieve maximum growth.

At an initial cell density of 9-15 x 10^3 mI^{-1}, the cells will generally undergo an eightfold increase in cell number before entering the stationary phase. Subcultures established with a high inoculums rate (0.5 -2.5 x 10^5 cells ml^{-1}) show an increase in cell number during the incubation period to a range 1.4 x 106 mI^{-1} before entering the stationary phase. The normal incubation time of stock cultures is 21-28 days between subcultures although cloning may occur within 18-25 days. In cases in which the cells are in a very active state of division, the passage length may be reduced to 6-9 days. Cell cultures initiated at very low cell densities will not grow unless the medium is enriched with the metabolites necessary to grow single cells or a small population of cells.

METHODS OF TISSUE CULTURE STUDY

Process of biotechnology in tissue culture may be broadly placed in four groups.

Somatic Cell Fusion

This is achieved by inducing two or more pro top lasts to fuse and the fusion product is nurtured to produce a hybrid plant. Though, this phenomenon is of common occurrence, protoplasts of some plant species fail to fuse. Some scientists have shown the production of hybrids that cannot be produced by the conventional hybridization methods because of-sexual or physical incompatibility among certain plants.

Genetic Engineering

Experiments are being conducted by some scientists to introduce neclei, chloroplasts, viruses, DNA, mitochondria, plastids etc. keeping in view the capability of isolated protoplast to ingest "foreign" bodies by a process similar to the process of endocytosis often seen in cells of protozoan and other animals.

Wall biosynthesis

Scientist have made use of isolated protoplast for the study of wall biosynthesis and deposition because of the ability of cultured protoplasts to regenerate a cell wall rapidly.

Study of protoplast population

As the protoplasts can be manipulated in a manner similar to that of the micro-organism, selection of mutant cell lines and the cloning of cell population is possible by using microbiological methods.

APPLICATION OF TISSUE CULTURE

In the recent years plant protoplast, cell and tissue cultures have become an important tool for crop improvement, commercial production of natural compounds and in the development of forestry. Some of the areas of biotechnological application of cultured plant protoplasts/cells/tissues are: tissue culture applications in order to capitalise on the totipotency of cells, cells and protoplast cultures coupled 'with DNA vectors to overcome problems caused by barriers to gene transfer through sexual means, culture of plant cells for the production of useful compounds, extension and increase of deficiency of biological nitrogen fixation, and transfer of genes for nitrogen fixation ability to non-fixing species. Due to the manipulation of plant tissues in the laboratory this technique has been referred to by some researchers, as a 'botanical laser' whose numerous uses are yet to be fully understood.

Application in agriculture

The major areas of application of tissue culture are briefly highlighted here.

- *Improvement of hybrids:* Development of cell fusion and hybridization techniques have solved the problem of incompatibility of plants and widened the scope of production of new varieties with in a short time. For example breeders obtained somatic hybrid of wild and cultivated potatoes (*S. tuberosum, S. chacoense*) and succeeded in the induction or organogenesis.

  The somatic hybrid plant inherited many characters viz., intermediate leaf morphology, stomata, forms and colour of tubers prolonged flowering, large and fertile pollen grains, high yield, resistance against Y-virus. However, haploid plant materials available as protoplast, cell and tissue culture systems are currently being evaluated for the use in transfer of foreign genetic material to selected plant species by protoplast fusion, transformation, transduction and organelle transfer.
- *Production of encapsulated seeds:* The two 'techniques, somatic embryogenesis and organogenesis are the alternative methods

for regeneration of the whole plant from cultured tissue in vitro. Moreover, the recent works on the wrapping an in vitro derived embryos in a seed coat ie. production of synthetic seeds are important because it may improve agriculture due to its low cost.

- *Production of disease resistant plants:* Many plant species, which propagate vegetatively are systematically infected by virus, bacteria, fungi and nematodes. Their inoculum is carried over several generations resulting in continued adverse effect of productivity and quality of crops. In order to ensure highest possible yield and quality, it is necessary to provide disease free stock plants to growers. Tissue culture techniques have solved this problem and minimized the time of biological testing.

Unless large scale population of pure inoculum of test pathogens are available, it is difficult to persue the establishment of pathogenicity and crop loss-assessment as it is done in field conditions. Now it has become possible to carry out such experiments in laboratory within short span of time by using tissue culture technologies. Scientists have discussed the following advantages for the study of several aspects of host-pathogen interactions and responses. They are:

- Ability to isolate host cells without wounding.
- Control of inoculum of pathogen and number of host cells.
- Ability t'J change the nature of host pathogen interaction by altering the constituent of growth medium.
- Presence of only one or a few major host cell types.
- Easy to apply and remove materials e.g. labeled precursors form cultured cells.

Production of stress resistance plant

Biochemical mechanisms exist is Cu1turetl cells which determine the resistance to biocide chemicals and provide the theoretical promise for selection in vitro. For example: cell suspension tolerant to 2,4-D when the cells were subcultured for six months in liquid medium supplemented with increasing amount of herbicides. The cells were able to grow in one rnM (milli mol) 2,4-D, while the control suspension was completely inhibited at O.3mM 2,4-D. It is suggested that tolerance was the result of induction of enzymatic systems responsible for the degradation of 2,4-D.

Uses of protoplasts as a system to select cell lines tolerant to herbicides have not been extensively explored. Protoplast technology can be used to increase the possibility to obtain monoclonal lines and offer the opportunity for intraspecific\transfer of cytoplasmic factor of resistance to some type of herbicide. A potential application of transfer of herbicides tolerance factor is the transplant of cytoplasmic organelles, for l example chloroplasts.

Transfer of nif gene to eukaryotes

Nitrogen fixing ability, a genetic character, exists in prokaryotic diazotrophs. However, one of the major tasks is the transfer of this character to eukaryotes. In recent years, researches are being done to solve this problem through tissue culture techniques coupled with the recombinants DNA tochnology . Historically, nitrogen fixation by rhizobia was believed to occur through symbiosis. For the first time an excitement was caused in scientific community with the discovery by Holsten et al. (1971). They obtained active rhizobia in the absence of nodules, leghaemoglobin and bacteroids which were apparently necessary in the intact plants. They established *Rhizobium joponicum* .In cell suspension of soybean roots. Callus induced from root explants of soybean on a specific medium was inoculated with R. joponicum. Later on it was microscopically observed that infection threads were formed by the bacteria which were present between intracellular spaces. They multiplied inside cells. Moreover, development of nitrogenase in soybean callus- Rhizobium system growing on solid was also observed and it was also found that only specific (isomorphic) cells of callus are vulnerable to infection by the bacteria.

In addition to improvement the bacterial strains and increased nodulation, it is necessary to seek those genotypes with the efficient photosynthesis and improved partitioning of carbohydrates to nodules.

Future Prospects

In addition to work done successfully on nif gene transfer, there are other important genes which have been clones, for example, (a) phaseolin and leghaemoglobin genes in soyabean; (b) storage protein genes in soybean, (c) genes of ribulose bi-phosphate carboxylase/oxygenase (Ru BP case) of pea, maize, wheat etc. Success achieved on these aspects would certainly promote in green revolution. Moreover, improvement in primary productivity by conversion of C3 plants into C4 ones through genetic engineering techniques hopefully would increase the primary productivity.

Application in horticulture and forestry

Tissue culture is applied in a number of cases in horticulture and forestry.

Micropropagation

Microprogation is a perfect alternative to asexual propagation of important plants and trees, because of following reasons.

- In this method only a small piece of tissue is needed to generate millions of clonal plants a year
- This method provides a possible alternative method for developing resistance in many species.

- It solved problem of quarantine for introduction of art disease in new areas and provide a mean for international exhange of plant material.
- By this method multiplication can be made in any season.

Regeneration of plantlets from cultured plant cells and tissues has been achieved in many trees of high economic value. Many of the studies are aimed at large scale micropropagation of important trees yielding fuel, pulp, timber, oils or fruits. Therefore, clonal forestry and horticultural are gaining an increasing recognition as an alternative for tree improvement. However, strategies for transferring cultured plants in vitro to field conditions are based on relatively higher priced horticultural species rather than agricultural and forestry species.

At planting out stage the plantlets fail to survive because of sudden change in the environment and invasion by soil microbes. So the regenerates should be transferred first to green house and then to field. The humidity should be controlled by covering the plants with transparent polythylene sheets. Some important plants on which work is done are: Acacia nilotica, Albizia lebback, Aprocera, Azadirachta indica, Bauhinia purpurea, Butea monosperma, Dalbergia spp, Dendrocalmus strictus, Eucalyptus spp, Ficus religiosa, Morus spp, Populus spp, Shorea robusta, Tectona grandis, Cryptomeria japonica, Picea smithiana and Pinus spp (all gymnosperms).

In vitro establishment of mycon-biza

Mycorrhizal fungi show highest specialization of parasitism. But major problems with them is their failure to grow on an artificial medium in laboratory. Therefore, establishment and multiplication of mycorrhizal fungi on cultured tissue of the same host plant, if successfully developed, may be a good tool for handling mycorrhizal fungi, production of high potential inoculum and their establishment in root systems of nursery plants in horticulture and forestry and the plantation of mycorrhiza infested seedlings into the field. Many attempts have been made to establish vesicul arbuscular mycorrhizal (VAM) fungi in axenic culture but unfortunately none of them got success. Moreover, micorhizal fungi have been cultured only on cortical tissues of roots which were seperated from the whole plant, as in root organ cultures where it acted as food base. VAM fungi have very high degree of specialization for food base on root cortex.

Application in industries

Production of useful compounds by cultured plant cells has become a field of special interest in various biotechnological programmes. However, much attention has been paid on the production of pharmaceutical and other secondary compounds such as essential oils, food flavourings and colourings which are used- by the fast food, ice cream and confectionary industries. This can be achieved by selection of specific cells producing high amount of desired compounds and the development of a suitable medium. In general, secondary metabolites produced by plant cell cultures are rather in small amount but strains of cells producing the same are in greater amount than those found in the intact plants have been isolated by clonal selection. Commonly two methods are employed for the selection of specific cells: Single cell cloning and cell aggregate cloning. The difficulties associated with isolation and culture of single cells limit application of this method. The later may appear to be more time-taking but easier than the first one.

Recently, in vitro production of high amount of useful compounds has increased with the success obtained so far in experimental studies. It is hoped that in near future, the industrial production of such compounds by using these techniques would be possible. Similarly, monogenic cultures of nematodes have been used for the study of mechanisms of action of nematicide. This technique can be used extensively in industry to supply nematodes for nematicide screaning programmes.

Area availability and Prime Institutions

Tissue culture/biotechnology in India is a fast developing science and there are areas it can be employed. Table 2 presents such areas and the products in which tissue culture can be applied. Table 3 enumerates the institutions which are conducting tissue culture and other biotechnological researches are being done on specific crop.

Table. Area of biotechnology/ tissue culture

Sl.No	Area of interest	Products
1	Recombinant DNA technology	Fine chemicals, enzymes (Genetic engineering) vaccines, growth hormones antibiotics, interferon
2	Treatment and utilization of biomaterials (biomass)	Single cell protein and mycoprotein, alcohols and biofuels
3	Plant & animal cell culture essential oils, dyes	Fine chemicals (alkaloids, steroids) somatic embryos encapsulated seeds, interferon, monoclonal antibodies
4	Nitrogen fixation	Microbial inoculants (biofertilisers)
5	Biofuels (bioenergy)	Hydrogen (via photolysis) alcohols (from biomass) methane (biogas) (from wastes and aquatic weeds)
6	Enzymes (biocatalysts)	Fine chemicals, food processing biosensor, chemotherapy
7	Fermentation	Acids, enzymes, alcohols antibiotics, fine chemicals, vitamins toxins (biopesticides)
8	Process engineering	Effluent water recycling product extraction, novel reactor, harvesting

It is to be noted that all the products mentioned in table above are industrially important which table 3 highlights the impor- tance given to crop and agricultural improvement.

Table. Crop and availability of biotechnology/tissue culture

Sl.No	Classification	Name of the crop	Technology available at
1	Fruits	1. Citrus	NBRI Lucknow, University of Jodhpur, Jodhpur
		2.Banana	IIHR Bangalore, NCL Poona
		3.Pineapple	BARC, Bombay
		4.Pomegranate	NCL, Poona
		5.Papaya	IARI, New Delhi
2	Species	6.Cardamom	NCL, Poona
		7.Ginger	NCL, Poona
		Turmeric	NCL, Poona, CPCRI, Kasargod
3	Ornamentals	9.Orchids	IIHR, Bangalore, NBRI , Lucknow
		10.Carnation	Delhi University, Delhi
		Gladiolus	NBRI, Lucknow; NCL, Poona; IIHR, Bangalore
		Bougainvillea	NBRI Lucknow, IIHR Bangalore
		Chrysanthemum	NBRI, Lucknow; NCL Poona
		Ferns	Baroda University, Baroda
4	Forest trees	Teak	NCL Poona
		Eucalyptus	NCL Poona, IISC, Bangalore
		Sandalwood	BARC, Bombay; IISC, Bangalore
5	Commercial crops	Sugarcane	NCL, Poona; SBI Coimbatore
6	Medicinal plants	Glycyrrhiza	GAU,Anand
		Jojoba	NBRI, Lucknow
		Dioscorea sp.	NBRI, Lucknow, RRL, Jammu

A brief idea of tissue culture one of the methods employed in biotechnology along with its scope and importance is given in this booklet. Indeed it is a highly specialized skill which pre supposes the knowledge of various biological sciences. Those who are desirous of getting more information could consult table 3 to find out the specific institution which are doing research on a specific crop.

PLANT TISSUE CULTURE

Plant tissue culture is a collection of techniques used to maintain or grow plant cells, tissues or organs under sterile conditions on a nutrient culture medium of known composition. Plant tissue culture is widely used to produce clones of a plant in a method known as micropropagation. Different techniques in plant tissue culture may offer certain advantages over traditional methods of propagation, including:

- The production of exact copies of plants that produce particularly good flowers, fruits, or have other desirable traits.
- To quickly produce mature plants.

- The production of multiples of plants in the absence of seeds or necessary pollinators to produce seeds.
- The regeneration of whole plants from plant cells that have been genetically modified.
- The production of plants in sterile containers that allows them to be moved with greatly reduced chances of transmitting diseases, pests, and pathogens.
- The production of plants from seeds that otherwise have very low chances of germinating and growing, i.e.: orchids and *Nepenthes*.
- To clean particular plants of viral and other infections and to quickly multiply these plants as 'cleaned stock' for horticulture and agriculture.

Plant tissue culture relies on the fact that many plant cells have the ability to regenerate a whole plant (totipotency). Single cells, plant cells without cell walls (protoplasts), pieces of leaves, stems or roots can often be used to generate a new plant on culture media given the required nutrients and plant hormones.

Techniques

Modern plant tissue culture is performed under aseptic conditions under HEPA filtered air provided by a laminar flow cabinet. Living plant materials from the environment are naturally contaminated on their surfaces (and sometimes interiors) with microorganisms, so surface sterilization of starting material (explants) in chemical solutions (usually alcohol and sodium or calcium hypochlorite or mercuric chloride[1] is required. Mercuric chloride is seldom used as a plant sterilant today, unless other sterilizing agents are found to be ineffective, as it is dangerous to use, and is difficult to dispose of. Explants are then usually placed on the surface of a solid culture medium, but are sometimes placed directly into a liquid medium, particularly when cell suspension cultures are desired. Solid and liquid media are generally composed of inorganic salts plus a few organic nutrients, vitamins and plant hormones. Solid media are prepared from liquid media with the addition of a gelling agent, usually purified agar.

The composition of the medium, particularly the plant hormones and the nitrogen source (nitrate versus ammonium salts or amino acids) have profound effects on the morphology of the tissues that grow from the initial explant. For example, an excess of auxin will often result in a proliferation of roots, while an excess of cytokinin may yield shoots. A balance of both auxin and cytokinin will often produce an unorganised growth of cells, or callus, but the morphology of the outgrowth will depend on the plant species as well as the medium composition. As cultures grow, pieces are typically sliced off and transferred to new media (subcultured) to allow for growth or to alter the morphology of the culture. The skill and experience of the tissue culturist are important in judging which pieces to culture and which to discard. As shoots emerge from a culture, they may be sliced off and rooted with auxin to produce plantlets which, when mature, can be transferred to potting soil for further growth in the greenhouse as normal plants.[2]

CHOICE OF EXPLANT

The tissue obtained from a plant to be cultured is called an explant based on work with certain model systems particularly tobacco it has often been claimed that a totipotent explant can be grown from any part of the plant and may include portions of shoots, leaves, stems, flowers, roots and single, undifferentiated cells.,[*citation needed*] however this has not been true for all plants.[3] In many species explants of various organs vary in their rates of growth and regeneration, while some do not grow at all. The choice of explant material also determines if the plantlets developed via tissue culture are haploid or diploid. Also the risk of microbial contamination is increased with inappropriate explants.

The specific differences in the regeneration potential of different organs and explants have various explanations. The significant factors include differences in the stage of the cells in the cell cycle, the availability of or ability to transport endogenous growth regulators, and the metabolic capabilities of the cells. The most commonly used tissue explants are the meristematic ends of the plants like the stem tip, auxiliary bud tip and root tip. These tissues have high rates of cell division and either concentrate or produce required growth regulating substances including auxins and cytokinins.

The pathways through which whole plants are regenerated from cells and tissues or explants such as meristems broadly fall into three types:

- The method in which explants that include a meristem (viz. the shoot tips or nodes) are grown on appropriate media supplemented with plant growth regulators to induce proliferation of multiple shoots, followed by rooting of the excised shoots to regenerate whole plants,
- The method in which totipotency of cells is realized in the form of *de novo* organogenesis, either directly in the form of induction of shoot meristems on the explants or indirectly via a callus (unorganised mass of cells resulting from proliferation of cells of the explant) and plants are regenerated through induction of roots on the resultant shoots,
- Somatic embryogenesis, in which asexual adventive embryos (comparable to zygotic embryos in their structure and development) are induced directly on explants or indirectly through a callus phase.

The first method involving the meristems and induction of multiple shoots is the preferred method for the micropropagation industry since the risks of somaclonal variation (genetic variation induced in tissue culture) are minimal when compared to the other two methods. Somatic embryogenesis is a method that has the potential to be several times higher in multiplication rates and is amenable to handling in liquid culture systems like bioreactors.

Some explants, like the root tip, are hard to isolate and are contaminated with soil microflora that become problematic during the tissue culture process. Certain soil microflora can form tight associations with the root systems, or even grow within the root. Soil particles bound to roots are difficult to remove without injury to the roots that then allows microbial attack. These associated microflora will generally overgrow the tissue culture medium before there is significant growth of plant tissue.

Aerial (above soil) explants are also rich in undesirable microflora. However, they are more easily removed from the explant by gentle rinsing, and the remainder usually can be killed by surface sterilization. Most of the surface microflora do not form tight associations with the plant tissue. Such associations can usually be found by visual inspection as a mosaic, decolorization or localized necrosis on the surface of the explant. An alternative for obtaining uncontaminated explants is to take explants from seedlings which are aseptically grown from surface-sterilized seeds. The hard surface of the seed is less permeable to penetration of harsh surface sterilizing agents, such as hypochlorite, so the acceptable conditions of sterilization used for seeds can be much more stringent than for vegetative tissues.

Tissue cultured plants are clones. If the original mother plant used to produce the first explants is susceptible to a pathogen or environmental condition, the entire crop would be susceptible to the same problem. Conversely, any positive traits would remain within the line also.

Applications

Plant tissue culture is used widely in the plant sciences, forestry, and in horticulture. Applications include:

- The commercial production of plants used as potting, landscape, and florist subjects, which uses meristem and shoot culture to produce large numbers of identical individuals.
- To conserve rare or endangered plant species.[4]
- A plant breeder may use tissue culture to screen cells rather than plants for advantageous characters, e.g. herbicide resistance/ tolerance.
- Large-scale growth of plant cells in liquid culture in bioreactors for production of valuable compounds, like plant-derived secondary metabolites and recombinant proteinsused as biopharmaceuticals.[5]
- To cross distantly related species by protoplast fusion and regeneration of the novel hybrid.
- To cross-pollinate distantly related species and then tissue culture the resulting embryo which would otherwise normally die (Embryo Rescue).

- For production of doubled monoploid (dihaploid) plants from haploid cultures to achieve homozygous lines more rapidly in breeding programmes, usually by treatment withcolchicine which causes doubling of the chromosome number.
- As a tissue for transformation, followed by either short-term testing of genetic constructs or regeneration of transgenic plants.
- Certain techniques such as meristem tip culture can be used to produce clean plant material from virused stock, such as potatoes and many species of soft fruit.
- Production of identical sterile hybrid species can be obtained.

Laboratories

Although some growers and nurseries have their own labs for propagating plants by the technique of tissue culture, a number of independent laboratories provide custom propagation services. The Plant Tissue Culture Information Exchange lists many commercial tissue culture labs. Since plant tissue culture is a very labour intensive process, this would be an important factor in determining which plants would be commercially viable to propagate in a laboratory.

PLANT AND FOREST BIOTECHNOLOGY

There is considerable scope for application of Plant and Forest Biotechnology in the State. The productivity of certain crops has declined in the State due to elevated temperature and salinity associated with drought and repeated floods. These problems can be addressed by the application of biotechnological techniques. Forests occupy 57,183.57 sq km, i.e. about 37% of the total geographical area, and play an important role in meeting the economic needs of the people of the State. The tribals in the State depend mostly on forests for their sustenance. Decrease in forest area, apart from affecting the tribal population, also affects the environment. Forest biotechnology can play an important role in mitigating these problems.

TISSUE CULTURE

The plant micro-propagation technology allows production of a large number of plants in a relatively small growing area, and in a short time, with a high degree of clonal phenotypic uniformity and absence of disease. In Orissa, although the practice of using tissue culture derived plants in the farming process is at a nascent stage, it deserves to be promoted. By innovative implementation of this technology in farming sectors, ornamental plants such as Caladium, Spathiphyllum, Philodendron, and plants like ginger, potato, asparagus, bamboo, banana, pineapple, papaya, strawberry and sugarcane etc. can be mass propagated through tissue culture. In order to make this possible, laboratories with skilled personnel will have to be established. A research station will be engaged in the standardization of protocols for tissue culture samplings and also for plants of different varieties so as to use them in the genetic transformation process. Large scale hardening facilities will be required for better survival of the tissue culture seedlings.

Development of genetic markers

It is essential to develop genetic markers for identification of elite germplasm and cultivated varieties. Efforts will be made to develop molecular markers particularly for genotypes possessing important attributes like resistance to specific diseases, such as bacterial leaf blight, stem borer infestation and blast in rice and other crops. Some of the local land races are highly resistant to wilting, particularly in case of solanacious crops. Hence, it is essential to identify the genes of interest and to isolate them for genetic transformation work.

A few land races and wild species of rice possess resistance to salt and drought stress. Efforts will be made to locate these genes and to transfer them into cultivated varieties through wide hybridization followed by embryo rescue technique. Marker assisted selection approaches would be developed and used to evolve such stress-tolerant crop varieties. Desirable genes will be identified by in-situ hybridisation. Molecular characterization of nematode resistant genes in green gram, black gram and common vegetables will be taken up.

Improvement in yield and quality:

The yield levels of rice, pulses and oilseeds have more or less stabilized. There are two ways by which yield status can be improved: (i) exploitation of hybrid vigour, (ii) identification of QTLs (Quantitative Trait Loci) in different crops and their use in genetic transformation. Yield of vegetables and fruits also can be improved by developing hybrid varieties. It is essential to bring genetic improvement in the crops commonly grown by the tribal population like millet, rice, beans, amaranth etc. for enhanced nutritional quality. Genetic improvement of major food crops for increased starch and vitamin content will be taken up, particularly for rice, potato, green gram and black gram.

Metabolic engineering of plants:

The crop productivity in the non-coastal area of the State is severely reduced primarily due to two major reasons: drought and elevated temperature. Coastal belt productivity is often also reduced because of floods and salinity. In order to sustain optimum productivity in these regions, there is a need for development of transgenic plant species that can with-stand these conditions. Several scientific organizations are capable of carrying in-depth research activities in this line. A long-term plan is proposed to be set up for developing plant species resistant to drought, elevation of temperature, salinity and floods, employing biotechnological methods.

Restoration and improvement of forests

There is an urgent need to develop biotechnological tools for forest restoration that can help improve forest and forest products in the following areas:

- Genetic improvement of timber yielding crops like teak, sal, sisoo etc. and selection of desirable types and their molecular characterization.
- Propagation of forest trees through macro- and micro- propagation methods particularly for the timber yielding species.

- Yield improvement of non-timber species viz., oil-yielding trees like *Azadiracta indica* (neem) and *Madhuca latifolia* (mahua) through gene transfer.
- Mass propagation by tissue culture and molecular characterization and genetic improvement of different bamboo species. Bamboo is an important raw material for construction of houses in the rural areas and is called the poor man's timber. Two main species of bamboo i.e., *Dendrocalamus strictus* and *Bambosa arundinacea* are dominant in Orissa's forests. These species need immediate attention due to their overexploitation. Hence, identification and selection of plus clumps, their conservation and mass propagation needs to be taken up on priority basis.

Biotechnological work in plant species shall be taken up on the following lines:

- Exploitation, collection, establishment and evaluation of tree germplasm of species like *Tectona grandis*(teak), *Santnum album* (sandalwood), Petrocarpus santilinus(red sanders), *Dalbergia sisoo*(sisoo), *M. latifolia*(mahua), bamboo etc. with due emphasis on germplasm conservation.
- Selection of candidate plus tree in each case and conducting of provenance and progeny trial.
- Macro- and micro-propagation studies and standardization of propagation techniques and mass propagation of plus and elite trees for operational planting.

- Field trial of plants raised by tissue culture.
- Establishment of clonal seed orchard for mass production of improved seed.
- Data base storage and retrieval of tree germplasm.
- Exploration and collection of seeds of various provenance and plus trees of priority species from the areas of natural distribution. Selection based on fast growing nature, straight bole, resistance to diseases and pests, and good adaptability.
- Studies on the floral biology and breeding systems.
- Evolving vegetative propagation techniques including macro propagation methods and mass propagation by tissue culture.
- Establishment of germplasm/gene banks for at least mid-term storage

APPLICATIONS IN HORTICULTURE AND FORESTRY

MICROPROPAGATION

In recent years, the application of micropropagation techniques as an alternative mean of asexual propagation of important plants has increased the interest of workers in various field. The micropropagation techniques are preferred over the conventional asexual propagation methods because of the following reasons : In this method only a small amount of tissue is needed as the initial explant for regeneration of millions of clonal plants in a year, *(b)* this method provides a possible alternative method for developing resistance in many species; it provides a mean for international exchange of plant materials, hence the problem for introduction of disease can be solved in quarantine; *in vitro* stock can be quickly proliferated as it is not season dependent, and valuable germplasm can be stored for a long time (Hu and Want, 1983; Mascarenhas and Muralidharan, 1989).

Regeneration of plantlets in cultured plant cell and tissues has been achieved in many trees of high economic value. Many of the studies are aimed at large scale micropropagation of important trees yielding fuel, pulp, timber, oils or fruits. Therefore, clonal forestry and horticultureare gaining an increasing recognition as an alternative for tree improvement. However, strategies for transferring cultured plants from*in vitro* to field conditions are based on relatively higher priced horticultural species rather than agricultural and forestry species (Fossard, 1987). Regeneration of plantlets in cultured tissue has been described (Murashige, 1974) to be accomplished into 3 stages *(see* shoot culture)'. Fossard (1987) has given a detailed account of stages of micropropagation

At planting out stage the plantlets fail to survive because of sudden change in the environment and invasion by soil microbes. Jagannathan (1987) has described that the regenerates should be transferred first to green house and then to field. The humidity should be controlled by covering the plants with transparent polyethylene sheets. This acclimatization requires several weeks whichshould be followed by potting into sterile peat or soil

In recent years, the interest has, aroused in commercializing the *in vitro* propagation of forest trees. This will bring about refinement in the existing procedures to make micropropagation more cost effective. Mascarenhas, Muralidharan and coworkers are making efforts to commercialize this biotechnology with respect to forest trees. However, development of automated procedure, plant delivery systems using somatic embryos and artificial seeds are also in progress. For betterment and improvement of tree plants of high economic value a break through in forestry research has come with production of artificial seeds in *Eucalyptus* (Muralidharan and Mascarenhas, 1989), and genetic transformation and *in vitro* regeneration in conifers (Gupta, 1989). Moreover, micropropagation has been successfully done in many trees (Gupta *et al.* 1980, 1981; Jaiswal and Pratap Narayan, 1985; Amin and Jaiswal, 1988; Mascarenhas and Muralidharan, 1989).

Mascarenhas and Muralidharan (1989) reviewed the tissue culture studies carried out on forest trees in India. Some of the important plants are : Acacia nilotica, Albizia lebbeck, A. procera, Azadirachta indica, Bauhinia purpurea, Butea monosperma, Dalbergia sp., Dendrocalmusstrictus, Eucalyptus sp. Ficus religiosa, Morus sp., Populus sp., Shorea robusta, Tectona grandis (all angiosperms), Biota oriental's, Cedrus deodara, Cryptomena japonica, Picea smithiana, Pinus sp. (all gymnosperms)

In Vitro Establishment of Mycorrhiza

Mycorrhizal fungi show highest level of specialization of parasitism. But the major problems with them is their failure to grow on an artificial medium in laboratory. Therefore, establishment and multiplication of mycorrhizal fungi on cultured tissue of the same host plant, if successfully developed, may be a good tool for handling mycorrhizal fungi, production of high potential inoculum and their establishment in root systems of nursery plants in horticulture and forestry, and plantation of mycorrhiza-infested seedlings into field. Only one report is available on this work. Kiernan *et al.* (1984) successfully produced strawberry plants by tissue culture which was infected by a mycorrhizal fungus, *Glomus* sp.

Many attempts have been made to establish Vesicular Arbuscular Mycorrhizal (VAM) fungi in axenic culture but unfortunately none of them got success. It was assumed that self inhibition of hyphal growth occurs in the growing germ tubes and the self inhibition compounds were recovered by adding activated charcoal into an agar medium that absorbs inhibitory compounds produced by germ tubes into medium. Cultures of mycorrhizas synthesized aseptically are grouped into two : the whole plant cultures and excised root cultures. Both the types of cultures are known as genetobiotic or monogenic systems (Rhodes, 1983). Due to presence of two organisms, it is also known as two member culture.

Mosse (1962) for the first time, reported the establishment of two member cultures. Appressorium formation and root penetration were much more likely to occur if a *Pseudomonas* sp. was present in culture. It is, therefore, suggested that 3 organisms *i.e.* fungus-plant-bacterium might be necessary for the development of symbiosis. Moreover, mycorrhizal fungi have been cultured only on cortical tissues of roots which were separated from the whole plant, as in root organ cultures where it acted as food base. VAM fungi have very high degree of specialization for food base on root cortex (Rhodes, 1983).

6

Modern Position of Tree Development in Forestry

ELEMENTS OF A GENETIC IMPROVEMENT PROGRAMME

The general objective of a genetic improvement programme should be the *sustainable* management of genetic variation to generate, identify and multiply for operational planting high-yielding and well adapted genotypes. For an outcrossing species, breeding typically incorporates:

- Establishment of an initial base population. In its broadest sense, for forest species this includes species and provenance testing, and the development of breeding and gene conservation populations.
- Population improvement. For forest species this typically includes recurrent cycles of selection and recombination.
- The derivation and multiplication of strains (e.g. families or clones) to be used operationally. This is the step in which breeding gains are captured and transferred to production populations.

The above are incorporated into a breeding strategy. A breeding strategy should permit continued improvement, and must allow for future alterations in breeding objectives. A sound breeding strategy also incorporates the early collection of appropriate biological and genetic data concerning the species, and should include measures to maintain a broad genetic base and to minimize inbreeding.

There exists a virtually unlimited number of alternative breeding strategies, varying with respect to, for example:

- Number of populations used
- Approach to selection (e.g. phenotypic, progeny test, clonal test, family)
- Mating design for recombination and the estimation of genetic parameters (e.g. open-pollinated, pair mating, factorials, diallels)
- Approach to handling the problem of minimization of inbreeding (e.g. pedigree control and sublining)

- Approach to the capture of genetic gain (e.g. open pollinated general combiner orchards, full-sib family options, clones)
- Method for integrating all of the above, and for balancing the frequently conflicting objectives - e.g. the problem of maximizing selection intensity while maintaining variation.

Timing of the different phases is an important component of a breeding strategy. Thorough species and provenance testing and biological studies are important early objectives. The desirability of early availability of improved genotypes for planting dictates, however, that some selection work will precede completion of the above. Some plantation programmes for certain species may not warrant the expense of an intensive breeding programme, and a genetic improvement programme may include little more than species testing and development of appropriate propagation methods.

THE STATUS OF TREE IMPROVEMENT

It has been estimated recently that, of a total of at least 50 000 tree species, some 400 have been the subject of formal breeding or testing activities of some description, 140 with at least one generation of selection and mating, and 35 of these involving at least 20 seed parents (Committee on Managing Global Genetic Resources 1991). The following is a brief survey of the current status of improvement for a representative sample of these species.

Industrial species

For convenience, the term "industrial" is applied here to species used predominantly in large scale plantations aimed mainly at the commercial production of timber. There are currently about 100 million hectares of industrial plantations worldwide (Gauthier 1991), and it has been estimated that projected demands for industrial timber early next century might be met by a total estate of 100–200 million hectares (Mather 1990). For convenience, industrial plantation species are grouped into broad classes in the following survey. It is admitted though that some groups are quite heterogeneous, and programmes for many species straddle climatic and rotation length classes.

Tropical and sub-tropical eucalypts

These currently comprise less than 5% of established industrial plantings, but this is a group of rapidly increasing importance due to the potential which exists for expanded plantings in many tropical countries. The most important taxa are *E. grandis, E. tereticornis* and *E. camaldulensis,* and notably hybrids between *E. grandis* and *E. urophylla* and the other two species. *E. grandis*is widely planted in moister areas, particularly of South America and Africa, due to its fast growth, good silvicultural characteristics, and good pulping qualities. *E. camaldulensis* and *E. tereticornis,* more drought hardy than grandis but without the same ability to capture the site, are very widely planted, particularly in India and Africa (Vivekanandan 1985, Koyo 1989). Other species with potential in tropical Africa, Asia and the Pacific, but much less widely used, include *E. microcorys, E. pilularis, E. cloeziana, E. citriodora, E. pellita,* and possibly *E. deglupta*(Burley & Barnes 1989). Private companies have featured prominently in the more well known plantings, but government enterprises are also important.

By the standards of most industrial forest plantation species, the tropical eucalypts are relatively easy to breed:

- Many flower abundantly and precociously
- They are generally characterised by short rotations, permitting early selection and short generations (e.g. generations of 4–7 years for *E. grandis*)
- Although not as easy as for pines, controlled pollination is still quite achievable
- Most coppice, and can be propagated reasonably easily by cuttings
- Seeds can be stored satisfactorily.

A few very advanced breeding programmes exist, undertaken by both government institutes and private growers. Four generations of selection and open-pollinated recombination have yielded large gains in the Florida (USA) programme (Squillace 1989). Spectacular and well-publicized successes have been recorded with clonal propagation of hybrids in Brazil and the Congo, although the breeding programmes underlying these clonal selection programmes are not as advanced (Campinhos & Ikemori 1989, Vigneron 1989). Clonal programmes are underway also for *E. grandis* in other tropical countries, but are perhaps unlikely to result in the large gains reported for the hybrids. These advanced programmes represent a very small proportion of the world plantation estate in tropical eucalypts. Provenance and progeny trials have been planted in many areas, some selections made, and seedling seed orchards established, but much of the planting is still conducted with material subjected to little if any improvement. Vigour, form and wood quality are typically the most important selection criteria. Cold tolerance has been an important criterion in the *E. grandis* programme in Florida (Meskimen 1983, Rockwood *et al.* 1989), although some freezes probably exceed the levels which can be tolerated by naturally available variants (Meskimen 1983). Most programmes are characterized by open pollinated approaches to recombination, testing and capture of genetic gain (e.g. Geary *et al.* 1983). Genetic variation of the tropical eucalypts is fairly well preserved in *ex situ* and *in situ* stands.

To summarize, genetic improvement programmes offer clear benefits for industrial plantation forestry with tropical eucalypts, and no major biological impediments to the breeding of these species exist. In terms of practical priorities, there is a lot to be gained by the extension of existing technologies to additional programmes. For the pure species, open-pollinated approaches are relatively simple and give good gains. Insufficient understanding and application of sound breeding strategies is a limitation in some programmes.

For the more advanced programmes, the availability of greater frost tolerance would be advantageous in some areas. The availability of better selection methods would circumvent the problem of the expense of clonal testing, and permit the imposition of higher selection intensities in the clonal programmes. Although generation intervals in these species are relatively short, large gains would be available through even earlier selection and flowering.

Cold-tolerant eucalypts

Although comprising no more than 2–3% of currently existing industrial plantations, this group is of expanding importance, particularly in plantation programmes by private companies in Chile, Africa and Europe. Around 50 000 ha of *Eucalyptus nitens* plantations have been established worldwide, while *E. globulus* plantations are much more extensive - over 1 000 000 ha in Portugal alone (Eldridge & Griffin 1990). Species for which plantings are more limited include *E. dunnii, E. smithii, E. radiata, E. fraxinoides, E. McArthurii* and *E. fastigata.*

Although generally grown on short rotations and selectable at a similar age to the tropical eucalypts, these species are generally more difficult subjects for breeding:

- flowering of E. globulus is poor in some locations, and E. nitens, E. dunnii and E. smithii are all shy flowerers
- these four species are difficult to root from cuttings.

E. globulus is characterized by sophisticated and well-planned breeding programmes which have been established in recent years, e.g. the "nucleus breeding" approach in Portugal, and the "multiple population" programmes in South Africa and Chile. Clonal and seedling seed orchards have been established for these species, but are yet to produce seed in required quantities. Selection criteria are principally vigour, stem form, and wood quality. Resistance to winter freezing is of some interest in relation to possible wider planting of *E. globulus* (Tibbets *et al.* 1991). Genetic resources of these species are relatively secure.

Addressing the flowering and rooting problems should be major priorities for more effective breeding of these species. Breeding programmes for these species would also benefit from the availability of greater frost tolerance, better selection methods, and even shorter generations, as described above.

Other tropical hardwoods

Although currently comprising less than 5% of established industrial plantings, this heterogeneous group is of rapidly increasing importance with considerable potential in many tropical countries. Some species, in particular

Tectona grandis, Gmelina and the tropical acacias, are already popular species in both private and public plantation operations.

Total plantings of *Gmelina arborea* probably stand at over 200 000 ha, with Brazil, West Africa, the Philippines and Malaysia the major centres. Potential exists in many other areas of the tropics, although site sensitivity is a limitation. The tree provides a very useful general purpose hardwood, and high quality pulp. With short rotations (7–12 years in some programmes), early flowering (at 3–5 years), reliable seed storage (under the right conditions), ready coppicing, and good rooting of cuttings, the species lends itself to rapid breeding, although only limited improvement has been undertaken. Controlled pollination is considerably more difficult than for the tropical eucalypts. Provenances have been quite widely tested, and seed orchards have been established in a few locations, e.g. Brazil, Malaysia, and the Philippines. Vegetative propagation of phenotypic selections is practised in Malaysia. Selection criteria are mainly stem straightness, cylindricity and volume.

Tectona grandis is widely planted in Indonesia (over 1 000 000 ha), Thailand, India, several other Asian countries, South and Central America and in many parts of Africa (where it is the most widely planted exotic) (White 1991). The species is moderately amenable to improvement - seedlings flower at age six to nine (age of first flowering is positively correlated with bole length to forking), yield of viable seed is often poor, seed can be stored, trees coppice, and juvenile cuttings root readily. Control pollination is feasible, but not easy. Long rotations, and a consequent long selection interval, is a limitation on rapid breeding. Trials in several places demonstrate wide provenance variation. Breeding programmes are in existence in Thailand, India and elsewhere, and many clonal seed orchards have been established (Burley and Barnes 1989, Kaosa-ard 1989, White 1991).

Acacia mangium has been planted extensively in Malaysia, and large plantings are continuing also in Indonesia. Other species such as *A. auriculiformis, A. crassicarpa* and *A. aulacocarpa* are also of interest. Seed production commences relatively early, seed can be stored satisfactorily, controlled pollination is feasible although not easy (Sedgley *et al.* 1992), and juvenile cuttings root well (Wong & Haines 1992). Broad provenance collections have been made and trials planted, and seedling seed orchards established at several centres. Gene pools are relatively secure.

Cordia alliodora is planted in Central America, Colombia, Venezuela and Ecuador, in plantations as well as an agroforestry crop (Newton *et al.* 1991). Rotations are about 20 years, and timber is of high quality. Heavy flowering commencing as early as two years and reliable seed storage (Greaves & McCarter 1990) are conducive to rapid breeding. Seed collections have been made and international provenance/progeny trials established. Seedling seed orchards have been established in Central America. The species is quite variable (Greaves and McCarter 1990), and some gene pools are endangered (Newton *et al.* 1991).

Triplochiton scleroxylon and *Terminalia superba* have been the subject of provenance/progeny tests and clonal trials in Africa, and plantations are being established. Selections are propagated by cuttings from stump coppice in the case of *Triplochiton,* and shoots from mature scions grafted onto young seedlings in the case of *Terminalia* (Leakey 1987). Rotation lengths and selection intervals are longer than for tropical eucalypts. *Paraserianthes falcataria* is planted in the Philippines and Malaysia, provenance/progeny trials have been established, and seedling seed orchards established. Gene conservation collections are being undertaken for *Bombacopsis quinata, Sterculia apetala, Alnus acuminata,* and *Vochysia hondurensis.* All have populations in danger of extinction or genetic erosion (Newton *et al.* 1991). Seedling seed orchards of *Bombacopsis quinata* have been established several countries in Central and South America including Honduras and Venezuela. Several members of the Meliaceae are potentially very valuable plantation species, e.g. *Swietenia macrophylla, S. humilis, S. mahogani* and *Cedrela odorata*. Damage by shoot borers, however, severely limits the extent to which plantations can be established. International provenance and progeny trials have been established of some of these species. Intraspecific variation in levels of resistance is evident, but the extent to which effective levels are available not determined (Newton *et al.* 1991, Newton *et al.* 1993). Some of these species have suffered severe genetic erosion. The use of valuable species of the Dipterocarpaceae (*Dipterocarpus elatus, Hopea* and *Shorea*) in plantations has been limited by flowering problems, seed recalcitrance, and silvicultural difficulties. Some of these species are also threatened by erosion of gene pools. A large number of other species, as yet untested or poorly known, are of potential use, including species of *Octomeles, Cleistopholis, Pterocarpus and Canarium.*

To summarize then, some tropical hardwoods compare well with tropical eucalypts as subjects for breeding, while some are a little more difficult. In general though, breeding is not yet as advanced. Broader implementation of good improvement programmes is an important priority for these species. Some other species pose major problems: e.g. insect susceptibility in the Meliaceae, flowering and seed problems in Dipterocarpaceae. Many other potentially valuable species are simply not well known, in terms of their adaptation and biological and genetic features, and gene pools are probably under threat in parts of the species ranges. The testing of potentially useful species, characterization of mating systems, provenance collection, establishment of trials, implementation of gene conservation measures and commencement of other breeding work poses a very large and urgent task.

Poplars and willows

Comprising perhaps 5–10% of established industrial plantations, poplars and willows are important plantation trees in many countries, including in

Europe, North America, China, Argentina, India, Africa, and Australia (Pryor 1992). Small private plantings feature strongly. The group includes a large number of species and subspecies.

Relatively short rotations, early and prolific flowering and ease of vegetative propagation are features which render the poplars amenable to rapid breeding and good capture of genetic gain, although seed storage is a problem. Apart from the long history of improvement, two features of poplar breeding programmes distinguish them from those for most other forest tree species (and present similarities to some fruit tree breeding programmes):

- The emphasis on between and within species hybridization, frequently involving well characterized clones, followed by clonal testing (Ostry & McNabb 1986, Carter *et al.* 1988, Herpka 1987). Most commercial plantings comprise tested clones, frequently with cultivar names as for ornamental horticulture.
- The emphasis on disease resistance (e.g. to *Melampsora, Marssonina* and *Septoria*) as the major selection criterion (Herpka 1987, Ostry & McNabb 1986, McNabb *et al.* 1990, Pryor 1992), with maintenance of growth characteristics a secondary objective. Disease resistance breeding has emerged in response to several clone-specific calamities.

The importance of the disease problem in poplars, and the long history of wide deployment of individual clones, may not be coincidental, and may provide a lesson for more considered and cautious deployment in clonal programmes with other species. An important factor limiting improvement in many poplar programmes is the availability of required levels of disease resistance. As above, large benefits would accrue through the development of better selection methods and shorter generation intervals.

Other temperate hardwoods

Plantations of species of this heterogeneous group, located in particular in Europe, North America and China, comprise perhaps 5–10% of the current industrial forest plantation estate. Although plantings of long rotation, high value species such as *Juglans* and *Prunus* are currently insignificant in total extent, it has been suggested that small private plantings of these could assume greater importance in the future on disused agricultural land (Von Althen 1991). For these species, small breeding programmes, by European and North American government institutes and cooperatives, are in existence. The long generation interval (related to both the long selection interval and the long delay to flowering), difficulty in controlled pollination, intermittent and light seed crops, and seed recalcitrance (Rink & Stelzer 1982, Beineke 1982, Robinson & Overton 1989) are limitations to the breeding of walnut. Orchards established many years ago are only now producing seed. Vigour and stem form are important selection criteria, but wood quality is also of particular importance. High values of the product have led to some use of grafted plantations, and patented clones are available (Beineke 1982, Beineke 1989, Robinson & Overton 1989). Unmanaged exploitation has led to scarcity of the species in many areas of its natural range (Robinson & Overton 1989). Breeding programmes incorporating clonal testing are underway for *Prunus* in Europe. With restricted current plantings and long generation intervals, breeding of these species beyond selection and either seed orchard establishment or propagation of superior genotypes is not readily justified. This is a group for which effective early selection would be particularly valuable.

Robinia pseudoacacia is a leguminous tree of considerable value in fast growing biomass plantations. Plantings stand at over 3 250 000 ha, with large areas in China, South Korea, and Hungary (Keresztesi 1983, Hanover *et al.* 1991). A precocious flowerer, regular and abundant seeder, with seed readily stored, and easily propagated from shoot or root cuttings, the species is very amenable to breeding. Hungarian breeding programmes commenced several decades ago have been based on crossing (including with related species) and clonal selection, and many registered cultivars are available (Keresztesi 1983). Selection criteria are wood quality (in particular), frost tolerance, growth and form. The origin and genetic base of many plantings, however, are unknown. Breeding programmes incorporating provenance and progeny testing have been commenced in North America (Mebrahtu and Hanover 1989, Hanover *et al.* 1991). Greater frost resistance in the species would extend its range.

Acacia mearnsii is widely planted for bark (tannin), mining timber, poles and fuelwood especially in southern and eastern Africa, and China. Difficult to control pollinate and to propagate vegetatively, breeding programmes are based on the open-pollinated progeny trial/seeding seed orchard approach (Raymond 1987, Hillis 1989). Alnus glutinosa is a very precocious (as early as two to three years) and heavy seed producer. Provenance/progeny tests have been established and seed orchards have been planned (Carter *et al.* 1988). Betula species similarly flowerer precociously and abundantly, and greenhouse seed orchards in Finland are now producing seed (Carter *et al.* 1988). Species involved in these programmes are B. verrucosa and B. pubescens. A plus tree selection programme for Betula verrucosa commenced in Sweden in 1988, and progeny trials and clonal tests have been established (Danell 1991). Small breeding programmes are in existence, and orchards have been established, for a number of other North American hardwoods, e.g. *Platanus occidentalis, Fraxinus pennsylvanica, Liquidambar styraciflua, Quercus rubra and Liriodendron tulipifera* (Squillace 1989).

Medium rotation conifers

Species of this group make up 35–40% of the world plantation estate, and potential exists for further plantings in many developing countries. Included are the important *Pinus* species *P. taeda, P. elliottii, P. caribaea, P. radiata* and *P. pinaster. P. taeda* comprises about 80% of some 10 000 000 ha of southern pine plantations in south-eastern USA (Lantz & Kraus 1987), and substantial areas exist also in China (220 000 ha) and Brazil. *P. elliottii* makes up the major part of the remainder of the USA plantations, and is important also in China (880 000 ha), Brazil (several hundred thousand hectares at least) and South Africa (some 150 000 ha). Substantial plantations of P. radiata exist in Chile, New Zealand and Australia (each more or less 1 000 000 ha). *P. pinaster* is an important plantation species in France (1 200 000 ha) and Portugal (900 000 ha), and is planted also in Chile and Australia (Destremau *et al.* 1982). World plantings of *P. caribaea* are over 500 000 ha, mostly in Venezuela although the species is important also in Australia, Fiji and Brazil, and potentially so in many other tropical countries. Several other pine species are important regionally, e.g. *P. patula, P. rigida, P. palustris, P. echinata, P. virginiana,* and *P. kesiya*. Some other pine species, e.g. *P. oocarpa* and *P. tecunumanii,* have potential in tropical regions.

Although the seed development cycle is longer than for many hardwoods, most of these pine species have some features which facilitate breeding:

- many flower reliably and well (although, for example, *P. caribaea* and *P. merkusii* flower poorly in some areas to which they have been introduced)
- controlled pollination is relatively easy
- seed can be stored readily
- grafting is easy

Cuttings can be propagated from young seedlings of many species, but propagation of cuttings from older trees generally is difficult.

Well established breeding programmes are in existence, at or near the third generation of selections in the USA *P. taeda,* New Zealand *P. radiata,* and Queensland *P. caribaea* programmes. (NCSU- I.CTIP 1990, Shelbourne 1986, Kanowski & Nikles 1989). A substantial proportion of the world's pine plantings are conducted with genetically improved material derived from the breeding programmes. The ease of controlled pollination has led to its frequent use in mating programmes. The large USA breeding programmes are typified by sophisticated mating designs for recombination and testing, e.g. diallels and factorials. Pair mating, polycrossing, or open pollination is more common elsewhere. Selection criteria are mainly vigour, stem form and wood quality, with disease resistance of some interest also, particularly for *P. taeda, P. elliottii* and *P. radiata*. The most advanced programmes employ controlled pollination also for the capture of genetic gain - to produce superior full-sib families, sometimes with multiplication by cuttings. The latter applies in particular to *P. radiata* in New Zealand and Australia, and to the hybrid between*P. elliottii* and *P. caribaea* in Queensland. Open pollination, however, remains the most common approach to the capture of genetic gain, and pollen contamination in seed orchards (50% even in well isolated orchards according to Hodge *et al.* 1991), is undoubtedly a cause of substantial leakage of this genetic gain in many programmes. Major limitations to improvement are the long intervals to flowering and selection (resulting in a long generation interval), inaccurate selection, and the poor rooting of cuttings from older trees. Gene pools for most species are reasonably well preserved in *ex situ* plantings, although some provenances of the Central American pines are endangered. Also included in this group are *Cryptomeria japonica* and *Chamaecyparis obtusa,* which make up the greater part of the Japanese plantations of some 10 000 000 ha. Historically, the Japanese programmes have been based on the vegetative propagation of selected varieties, each comprising a small number of phenotypically similar clones. Provenance and progeny tests have been established and crossing programmes implemented in more recent times.

Araucaria cunninghamii is a valuable timber tree represented by 50 000 ha in Australia but with considerable potential elsewhere in the tropics. Mature trees coppice readily, and cuttings root well, but a rigid orthotropic and plagiotropic branching system places restrictions on vegetative multiplication rates with cuttings. Irregular flowering, and long delays to the production of male strobili, place further constraints on the breeding of the species, although an active breeding programme tailored to accommodate these characteristics has resulted in substantial gains (Nikles *et al.* 1988). Another fairly widely planted species of this genus is *A. angustifolia,* in South America. For this species, however, little improvement has been undertaken, and the species is endangered in parts of its natural range. Genetic improvement offers clear benefits for planting programmes with species of this group. Broader application of good improvement programmes is an important priority, particularly in tropical countries. The broader development of technologies for the mass production of superior full-sib families would also be highly beneficial. In terms of research objectives in advanced breeding programmes, major benefits are likely to accrue from more effective control of the maturation state (particularly in relation to the rooting of cuttings), more effective selection, and shorter generation intervals.

Long rotation temperate conifers

This group, comprises 35–40% of global industrial plantations. Important species included here are *Pinus sylvestris* (annual plantings of 90 000 ha, 90 000 ha, 100 000 ha, 40 000–50 000 ha in Finland, Sweden, China and Poland respectively, and with additional large programmes in countries of the former USSR), *Picea abies* (a major plantation species in Europe and North America), other *Picea* species, e.g. *P. sitchensis, P. mariana P. glauca* and *P. engelmannii* (important in U.K., Canada and north-eastern USA), *Pinus contorta* (at least 350 000 ha established in Sweden (Fries 1987), and large areas also in Canada (Ying *et al.* 1985, Konishi 1986)), *Pseudotsuga menziesii* (a very important plantation species in north-western USA and Canada, but represented by some 350 000 ha also in Europe), *Pinus banksiana* (important in eastern Canada), *P.*

strobus (USA), *Larix* species (extensive plantings in South Korea and grown also in north-eastern USA and Europe). Some large planting programmes are undertaken by government bodies, e.g. in the UK and Canada. A prominent feature of the industry in several European countries is the large number of small growers. *Pinus contorta* (Fries 1987) and *P. banksiana* (Park *et al.* 1989) are reasonably precocious and reliable flowerers, but several of the other species mentioned are characterized by light and/or irregular flowering, e.g. *P. sylvestris* (Matyas 1991, Koski 1991), *Picea abies* (Wellendorf 1989, Dietrichson 1989, Ruotsalainen and Nikkanen 1989) and *P. engelmannii* (Konishi 1986). The major obstacle to improvement, however, and the main difference from the medium rotation conifers, is the longer selection interval, rendering traditional breeding very slow.

Well established breeding programmes, largely the responsibility of government institutes, exist for these species. Most testing and recombination programmes are based on open-pollination (Lindgren 1991. Fries 1987, Morgenstern and Park 1991) although some pair mating is used, and more sophisticated mating designs for Douglas fir. Interspecific hybridization is of significance in *Larix* (Paques 1989). Some of these species display marked provenance X planting site interaction, particularly in relation to temperature sensitivity, and definition of breeding zones is an important feature of some programmes, e.g. for *Pinus sylvestris* (Danell 1991), *P. banksiana* (Miller 1984), and *Picea abies* (Dietrichson 1989). Major selection criteria are vigour, straightness and wood properties, frost resistance in *Pinus sylvestris* (Oleksyn 1991) and *Picea abies* (Van de Sype 1989), disease resistance in *P. contorta* (Ying *et al.* 1985), and insect resistance in*Picea abies.*

The open-pollinated orchard, of which large areas have been established, remains the main approach to the capture of genetic gain, but most orchards are not producing enough seed to make much impact on the requirement for planting stock. Approximately 200 ha of clonal tests have been established in Germany (Kleinschmit & Svolba 1989), and over 10 000 clones are under test in Sweden (Danell 1991). Commercial use of clones, however, remains very limited, due partly to a reluctance of growers to pay higher prices for cuttings (P. Monchaux pers. com.) and to doubts concerning the capacity to retain sufficient levels of juvenility through a lengthy sequential propagation programme (Kleinschmit & Svolba 1989). Most gene pools are reasonably well preserved. The broader development of technologies for the mass production of superior full-sib families would be useful for these species. Some programmes would benefit from a greater capacity to manipulate frost tolerance, and some from the ability to manipulate flowering. Breeding of the long rotation species would benefit particularly from reductions of the generation interval through early selection. It should be noted that in some countries, plantations of these species are used for many purposes. Intensive breeding for enhanced timber production in such plantation programmes may not be desirable, particularly where reduction of genetic diversity is involved. Legislation reinforces this view in some countries.

Non-industrial Species

There is a large requirement worldwide, mainly in developing countries, for fuelwood plantations, plantings to arrest and reverse land degradation, and, most importantly, tree plantings integrated into farming systems to provide mulch, fodder, shade, fuelwood, timber, fruit, and soil improvement and protection. Many of the species discussed in this category can be used in industrial plantations (as defined above), and identification of a "non-industrial" group of species is a demarcation which is more convenient than clearly defined. Large fuelwood plantations, e.g. of tropical eucalypts or casuarinas, located near cities are one approach to the urban fuelwood crisis. Operated on a commercial basis, these can be viewed as industrial species. Current plantings are not widely based on improved material, but simple improvement programmes would be justified in some of these.

Major rehabilitation plantings, e.g. with casuarinas or acacias, are publicly funded. Improvement beyond species and provenance selection may be difficult to justify on purely economic grounds, and may furthermore be of limited desirability in those cases where maintenance of diversity in such plantings is of major importance.

Examples of integration of trees into farming systems include:

- The alley farming systems, incorporating the use of trees such as *Gliricidia* or *Leucaena* for protection and mulch production.
- Shade tree systems, e.g. *Inga* and *Erythrina* species grown as shade over coffee in Latin America.
- Trees grown principally for fodder and fuelwood in agroforestry systems, e.g. *Prosopis* and arid zone acacias.
- Trees grown in association with crops mainly for timber, e.g. *Paulownia* in China and *Dalbergia* in Pakistan.
- windbreak plantings, e.g. of *Casuarina*.

Gliricidia is a fast growing tree used widely in the Philippines, India, Sri Lanka, Indonesia, East Africa and its native Central and South America for fodder, fuelwood, live fences, construction wood and mulch. Subjected to agricultural use for centuries, its true natural range is uncertain, and many land races have been created (Hughes 1987). Flowers are produced within the first year or two, seed production is prolific, and cuttings are easily propagated. Large collections have been made and some provenance comparisons carried out, but breeding is not very advanced. A breeding seed orchard of a particular desirable provenance will be established at the ICRAF research station in Zambia (Simons 1992b). Cuttings of very limited numbers of locally selected clones are the planting stock in some areas. Genetic variation with respect to a range of characters is substantial (Simons 1992b). Selection criteria vary with use, but timing and reliability of leaf production are important parameters for fodder production.

Leucaena is a long cultivated tree of importance in South East Asia, Nepal, India, Africa and Latin America. It is nitrogen fixer which provides good forage, green manure, firewood and small timber. Frequently used with other crops, it is also the major component of energy plantations established in the Philippines (Durst 1987). *L. leucocephala* seeds very precociously and prolifically, but is not easily propagated by cuttings. Unlike other tree species reviewed here, *L. leucocephala* is frequently self-pollinating, although this is not a feature of other species of the genus. With respect to genetic improvement, *Leucaena* has received more attention than most other non-industrial trees. Two directions have been followed in the Hawaiian programme of Dr Brewbaker:

- Testing of self-pollinated lines
- Interspecific hybridization

In most areas, the psyllid *Heteropsylla cubana* causes serious damage (McDicken & Taylor 1988, Durst 1987), and resistance to this pest is an important selection criterion. Others include biomass, cold tolerance, seedlessness, fodder quality, and dry season leaf retention. Although yield increases of well over 100% have been achieved through simple selection (MacDicken & Mehl 1990), breeding is yet to have a major impact in planting programmes, many of which have suffered from off-site planting and a very narrow genetic base. For *L. leucocephala*, locally produced seed is what is generally used. Some *Leucaena* species are under threat, and collections are underway.

Calliandra calothyrsus is a nitrogen fixing, thornless legume producing high quality fuelwood and fodder, with considerable potential in site amelioration, particularly sites too wet for *Gliricidia*and *Leucaena*. The species flowers in its second year and coppices and roots well. Provenance and progeny tests have been planned (Burley & Barnes 1989, MacQueen 1992). *Prosopis* is an arid zone genus of about 45 species, highly tolerant of drought and of difficult soils, grown widely on marginal land in India and countries in South America and with potential in many other countries. The trees provide fuelwood, construction timber and fodder. Coppicing is prolific (Walker 1988). A range of provenance trials coordinated *i a* by FAO, involving several species, are in existence (Cossalter *et al.* 1987, FAO 1988a, Matheson 1990, Dunsdon *et al.* 1991, Bach 1992, Graudal & Thomsen 1992). Objectives of breeding include growth increases, reduction of thorniness, and improvement of fodder quality.

A mimosoid legume genus of about 400 species, *Inga* is planted widely in Latin America to provide shade over coffee, wood, fuelwood, nitrogen fixation, green manure and fodder. Seed recalcitrance has limited plantings in other regions. Little collection work has been undertaken.

There are about 135 African acacias, of which *A. nilotica, A. tortilis, A. senegal,* and *Faidherbia alba* are particularly important. Very drought resistant, these species provide fodder, timber, charcoal, shade, honey, gum arabic, tannins and dune stabilization in arid areas of Africa and the Indian subcontinent. Some species display salt tolerance. The species tend to be prolific seeders, seeds can be stored, and stumps produce coppice (Fagg & Barnes 1990). Species are characterized by considerable genetic variation, and some collections have been undertaken and provenance trials established under the auspices of FAO (Cossalter *et al.* 1987, FAO 1988a, Koyo 1989, Bach 1992, Graudal & Thomsen 1992). *A. nilotica* is regarded as a weed in Indonesia and Australia.

Erythrina poeppigiana and *E. fusca* are used as pollarded trees providing shade and mulch for coffee in Central America. *E. berteroana* is used in living fences. The species flower precociously (at three years) and abundantly. Cuttings root well, and direct planting of unrooted stakes is common. Improvement work conducted at CATIE is along poplar lines, involving the clonal testing of individuals from natural and planted stands.

The genus *Paulownia* is represented by about nine species native to China, where there are reportedly 1 300 000 ha planted in mixed cultivation with crops (Chinese Acad. For. 1986). The species is suited to intercropping due to late leaf emergence and leaf fall, and a deep root system (Chinese Acad. Forestry 1986), and produces timber within five to six years. *P.tomentosa*and *P.kawakamii* usually flower in the second year after planting, and *P.fortunei* and *P. catalpifolia* in the fifth to sixth year. Seeds can be stored satisfactorily, and cuttings can be propagated from both young seedlings and the roots of mature trees. In China, the approach to breeding envisaged is controlled pollination (in particular interspecific hybridization) and then clonal testing (Chinese Acad. For. 1986). Selection criteria are height (to minimize shading of crops) and freedom from the many diseases and pests. Planting stock currently used comprises selected clones of some good hybrids. The North American breeding programme is at the species and provenance evaluation stage (NCSU-Industry Hardwood Research Co-op. 1990).

Dalbergia sissoo has long been valued in Pakistan, India and Nepal for its high value cabinet timber, fodder, fuelwood, charcoal, honey, medicinal properties, windbreak quality and capacity to fix nitrogen (White 1990). It is used both with crops and as an industrial plantation species (Sheikh 1988). The species flowers abundantly and regularly from age four, and good rooting of cuttings has led to their use as the most common method of propagation (Sheikh 1988, White 1990). The breeding system of the species is not well known (White 1990). Seed orchards were established in Pakistan in 1975 but improvement is not very advanced (White 1990). Selection criteria are wood yield and quality, but with consideration given also to fodder and honey properties.

Casuarina equisetifolia and *C. cunninghamii* are used widely for windbreak plantings and fuelwood production in areas of the tropics. *C. equisetifolia* comprises the major component of a shelterbelt covering over 1 000 000 ha in China (National Research Council 1984), Coastal sand dune fixation is one of the most successful applications of this species. Both species are dioecious. For industrial plantations, provenance trial establishment, plus tree selection, and the establishment of Breeding Seed Orchards using the multiple population approach has been recommended (Burley & Barnes 1989). Breeding of *Casuarina* in Egypt incorporates selection for drought resistance (Burley 1985). A sterile hybrid between *C.junghuhniana* and *C.equisetifolia*is propagated vegetatively and widely planted in Thailand (Willan *et al.* 1990).

Azadirachta indica is a promising species of uncertain natural distribution due to centuries of traditional use (Willan *et al.* 1990). Poor seed viability is a serious problem. Seed collections have been made and provenances are under test (F/FRED 1988, Nikles 1992). Growth rate, insect resistance and azadirachtin content will be important selection criteria (Nikles 1992). A large number of other species are of potential value in non-industrial plantings, most remaining poorly studied. These include species of *Sesbania, Pithecellobium, Albizia, Caesalpinia* and*Parkinsonia* (MacDicken & Mehl 1990, Dunsdon *et al.* 1991).

To summarize for the non-industrial species then, moderately comprehensive data on biological features influencing breeding are available for only a handful of species. Several of these taxa are highly variable and display features which render them particularly amenable to very rapid improvement by traditional means - precocious and abundant seeding, coppicing, and ease of vegetative propagation. Biological features of many potentially valuable non-industrial species remain largely unknown, and collection of data of this type will be an important component of the initial phase of any breeding programmes. Although not well studied, some gene pools are under threat, and some species display unpredictable patterns of variation as a result of human disturbance, rendering sampling strategies more difficult to design.

Considered collectively, non-industrial trees differ markedly from industrial trees in the wide variety of end uses and thus selection criteria - e.g. fuelwood quality, foliage yield, fodder quality, coppicing ability, phenological characteristics, timber quality, nitrogen fixation, root features, tree form and resistance to insects, disease, drought and other stresses. Some of these, e.g. fuelwood quality, are very complex. For some farmers, reliability of production may be more important than mean annual yield (Simons 1992a). In agroforestry systems, the interactive effect of the tree component with the associated crop is important (Willan *et al.* 1990). Desired features of agroforestry species have sometimes been integrated in terms of "ideotypes" (Glover 1990). Although selection work has been undertaken in a few programmes, and clonal propagation of selected clones is used in *Paulownia* and *Erythrina,* most non-industrial species remain at the species testing stage. The work involved in just selecting the most promising of perhaps thousands of other potential valuable but as yet untested species is formidable.

Improvement beyond species and provenance testing may not be warranted in many programmes with non-industrial species:

- the value of the product may not warrant the effort, particularly where the tree crop is the lesser component of the farm product.
- selection criteria for a particular species may vary considerably with farming system and region.
- there may be some difficulties in transferring the results of central breeding programmes to the growers.

Major emphases of improvement for non-industrial species are likely to be taxonomic studies of variation, species and provenance testing, the assessment of some features of reproductive biology, and conservation activities. Where regionally consistent selection criteria can be defined, some clonal selection may be important for species easily propagated vegetatively, and perhaps seed tree selection for species propagated by seed. Sophisticated breeding programmes are unlikely to be warranted for most of these species. Interspecific hybridization is of proven value for some taxa, and holds promise also for others, but must be used in the context of a sound improvement programme.

TREE IMPROVEMENT IN DEVELOPED COUNTRIES

Tree improvement in developed countries mainly concerns industrial species, although agroforestry is becoming more important. Some form of genetic improvement programme is a feature of most plantation programmes for industrial species. Most are based on recurrent cycles of phenotypic selection (frequently followed by testing of progeny) and recombination among selected individuals. Exceptions are the poplars and cryptomeria, for which clonal selection has been the traditional approach. Vigour, stem form and wood quality are the most important selection criteria in most programmes. Resistance to disease or insects in particular, and also to cold and drought, are significant in several programmes. The importance of disease resistance in the poplars is worthy of special note. Most characters of major importance are under polygenic control. Although there are some clonal programmes, and programmes based on the propagation of full-sib families, the open-pollinated orchard is the most common approach to the capture of genetic gain. Gene pools are reasonably well preserved for most of these species, although locally adapted populations may be in danger of depletion or genetic pollution in some cases.

Breeding activities (e.g. crossing programmes, progeny testing, selection) are generally carried out under the guidance of personnel with expertise in quantitative plant breeding, supported by good systems of field experimentation and data base management. The personnel involved are frequently employees of the grower organizations, although independent research institutes are involved in some cases. Breeding cooperatives, sharing genetic material, knowledge and work, are particularly important for *Pinus taeda, P. elliottii, P. radiata,* and some north American hardwoods (e.g. the North Carolina, Florida and North Central Hardwood cooperatives, and the Southern Tree Breeding Association).

Supportive research (e.g. assessment of reproductive behaviour, mating patterns, genetic parameters) is conducted frequently by the above groups, but also by universities and other research organizations. Strategic research (aimed at the development of better breeding methods) is undertaken primarily by universities and other research organizations, although several of the larger private growers also conduct long term strategic research. As pointed out by Kanowski (1993), research in tree improvement is greatly under-resourced, in both absolute terms and relative to agriculture. Significant genetic gains are being achieved in these breeding programmes but, in particular for the long rotation species, there has been only a minor impact to date on the genetic quality of plantations.

Major limitations to rapid improvement with most of these species are:

- Long generation intervals, related to:
- poor juvenile-mature correlations. Although there is some debate about this, in general, for many characters, these correlations are not high enough to permit very early selection. For *Pinus elliottii,* for example, it has been suggested that growth up to five years involves different physiological mechanisms to subsequent growth (Hodge *et al.* 1991). It should be noted that selection intervals for many of the breeding programmes discussed above are shorter than those recommended by some authorities. Zobel & Talbert (1984), for example, advocate half rotation age as the appropriate selection interval.
- long juvenile phase with respect to flowering.
- Low effectiveness of selection for many characters. Most major selection criteria are of low heritability, such that progeny testing is necessary. Some characters, e.g. resistance to frost, drought and insects, are difficult to test in a systematic way.
- Through the use of open-pollinated orchards, the exploitation of only a part of the genetic variation available.

Major research priorities for these species should be the broader development of methods for the propagation of full-sib families and clones, and the development of methods for early and more accurate selection and the promotion of precocious flowering. It is of some interest to note that an apparently slow rate of advancement with respect to genetic quality of plantations is not unique to forest tree species. The avocado varieties Fuerte and Hass, for example, were introduced to the industry over 60 years ago, but remain the major commercial varieties. Similarly, the major kiwifruit varieties were selected several decades ago. The advances which have taken place with these crops have been horticultural, not genetic. Silvicultural advances are critical also for forest tree species, but it is unlikely that the intensity of horticulture given to fruit tree crops would be affordable. For some agricultural species, the genetic superiority of modern varieties requires intensive agriculture, including comprehensive pest and disease management, for its expression. Modern high yielding varieties tend to lack much of the natural resistance of old land races or wild relatives (Harms 1992). The lower intensity of silviculture affordable renders desirable, then, the more careful management of genetic improvement for forest tree species.

TREE DEVELOPMENT IN RISING COUNTRIES

There are many examples of very successful industrial plantation programmes, including some of the world's best, in developing countries. Mostly these are managed by private forestry companies, and involve biologically well-known plantation species. These operations employ breeders, and in some cases hire consultants. Limitations and priorities for these programmes are much the same as those discussed above.

As noted above, an additional area of up to 100 million hectares of plantations will be required globally in order to meet future predicted demand for industrial timber. It is likely that the bulk of this additional area will be located in tropical countries. To some extent, this will be accomplished with already proven industrial species - conifers, tropical eucalypts, and other established tropical hardwoods. It is likely, though, that the use of other industrial species will be required:

- To utilise a broader range of sites
- To meet a demand for products currently provided by harvesting in natural forests.

Selection criteria for these "new" industrial species will be similar to those for established species. Some are probably very amenable to improvement, while others present problems, e.g. flowering and seed problems, and insect susceptibility. Many potentially valuable species are simply not well known, in terms of their adaptation and biological and genetic features, and gene pools are probably under threat.

There is a global requirement also for several hundred million hectares at least of non-industrial plantings, mainly located in developing countries. It is noteworthy that while the plantation area under industrial species in the tropics doubled in the 1980s, the area under non-industrial species almost tripled (FAO 1993). Non-industrial species differ markedly from industrial species in the wide variety, and the complexity, of selection criteria. Several of the taxa are highly variable and display features which render them particularly amenable to very rapid improvement by traditional means. Biological features of many potentially valuable non-industrial species, however, remain largely unknown. Although not well studied, some gene pools are under threat, and some species display unpredictable patterns of variation as a result of human disturbance. Although selection work has been undertaken in a few programmes, most non-industrial species remain at the species testing stage. There are some 2 000 tree species that have been planted as "multipurpose" trees (Burley 1985), and many more species would have potential. As pointed out by Wood (1991), the number of tree species of potential interest in developing countries is much larger than the number of food crops in these regions. Undoubtedly, the number of species to be covered limits the progress which can be made with any one. It is therefore necessary to address the question "to what extent is interest in a large number of species necessary?". While selection of an appropriate site is a critical factor in the establishment of any industrial plantation, the establishment of non-industrial plantations is often less flexible - plantings are frequently required in a particular area, e.g. for rehabilitation. Non-industrial species are thus required for a wide variety of sites and conditions. Furthermore, sites are often marginal, with variable environments and subject to periodic stresses, and testing over several years is desirable. Socio-economic considerations will furthermore strongly influence the choice of species for local use. For these reasons, reliance on a few species is unlikely to be feasible. The work involved in selecting the most promising species is formidable.

Genetic improvement of the new industrial species and non-industrials is generally the responsibility of state, national, and in particular regional and international organizations. Involved are international bodies such as FAO, agencies operating bilaterally, e.g. AIDAB, ACIAR, DANIDA, FINNIDA and GTZ, research organizations such as CAMCORE, CSIRO, CIRAD-Forêt, IBPGR, CIFOR, ICRAF, OFI, CATIE, and F/FRED, and many national institutes. Typically, limited resources are very thinly spread. Growers are unlikely to conduct formal breeding themselves, and frequently may prefer to minimize their expenditure on establishment by producing their own planting stock. General limitations to tree improvement in developing countries are a lack of staff with the relevant skills (Griffin & Nikles 1984, FAO 1988b, Koyo 1989, Mburu 1989, Brouard 1990, Kio 1991, de Freitas 1991, Ng 1991), inadequate facilities (FAO 1988b, Koyo 1989, Mburu 1989), poor communication, particularly with forest managers (Griffin & Nikles 1984, FAO 1988b) and poor language skills (Ng 1991). It is these resource limitations and the variability in user requirements, rather than biological limitations, which constitute the major impediment to the rapid improvement of non-industrial species (almost totally dependent on public funding).

The broader implementation of good improvement programmes will be an important priority for the new industrial species. The testing of potentially useful species, characterization of mating systems, provenance collection, establishment, management and evaluation of trials, implementation of gene conservation measures and commencement of other breeding work, poses a very large task.

As noted above, major emphases of improvement for non-industrial species are likely to be taxonomic studies of variation, species and provenance testing, the assessment of reproductive features, and conservation activities. Interspecific hybridization may be of value for several taxa. Where regionally consistent selection criteria can be defined and the tree crop is considered sufficiently important, clonal selection programmes are likely to be of value for species easily propagated vegetatively, and seed tree selection for species propagated by seed. Sophisticated breeding programmes are unlikely to be warranted for most species. Simons (1992a) makes the point that new genotypes may need to be markedly superior to existing material to be adopted by farmers. This argues for a once only improvement effort, or perhaps for breeding for occasional "quantum leaps" rather than incremental gains. Improvement programmes for non-industrial trees, as for industrial species can only be conducted effectively where planting programmes are based on sound establishment and management practices.

USES OF FOREST TREE SPECIES IN SOME SADC COUNTRIES

Both exotic and indigenous forest trees species that provide goods and services such as firewood, fruits, timber, poles, fodder, environmental protection, amenities in individual SADC countries were identified in the country reports.

TREES IN AGROFORESTRY SYSTEMS

Both indigenous and exotic tree species are used in agroforestry systems. The species are either deliberately planted or are managed within their natural habitat to improve their productivity. Among the most important indigenous tree species include the African acacias (*Acacia erioloba, A. karroo, A. nilotica, A. senegal, A. tortilis*), *Dichrostachys cineria, Faidherbia albida* and *Sesbania sesban.* Exotic agroforestry species that were identified as important species in some of the SADC countries include *Acacia angustissima, Calliandra calothyrsus, Gliricidia sepium, Leucaena leucocephala, L. pallida, L. diversifolia* and *Tephrosia vogelli.*

Trees for firewood

Although, practically all tree species could be used for firewood, there is a group of species that is generally preferred. Such species usually burn without excessive smoke and unpleasant odours. The choice of species is however now very restricted in some localities due to unavailability of some of the preferred species in sufficient quantities. Among the indigenous species, the most widely used tree species in the different SADC countries were identified as *Acacia erioloba, A. karroo,A. nilotica, B. plurijuga, Brachystegia* spp.,

C. mopane, *Combretum* spp., *Dialium schlehteri*, *F. albida*, *J. globiflora*, and *Terminalia* spp. Exotic tree species that are being grown for provision of firewood include *Eucalyptus* spp., *C. cunninghamiana* and *Acacia mearnsii*. Pines and cypresses although often used in areas of high wood deficit are not preferred species. Where, *A. mearnsii* is commercially grown for its tannin bark, its wood is also used in the commercial production of charcoal particularly in South Africa and Zimbabwe.

Trees used for food production

In all the country reports, forest tree species were mentioned as important sources of food from their fruits, more so in times of food shortage. Forest tree species also directly provide other edible products such as seed and nuts, gums and resins, root tubers and leaves for vegetables. Indigenous fruits, for example are processed into a variety of products that store well and therefore readily available in periods of food shortage (Campbell, 1987; Maghembe *et al.*, 1984). Some indigenous fruits such as *B. discolor*, *Phoenix reclinata*, *S. birrea*, *U. kirkiana* and *Z. mauritiana* are used in the production of alcoholic beverages. The Amarula cream, an alcoholic product developed from *S. birrea* fruit in South Africa is now marketed in about 50 countries (Ham and Van Eck, 1997). In Namibia, Hailwa (1999) estimated that fruits and beverages exploited annually are valued at N$4.8 million. Although there are more than a hundred indigenous tree species found in the SADC region that produce edible fruits the following species are regarded as the most important in this diverse region: A. digitata, Annona senegalensis, Artabotrys brachypetalus, Azanza garckeana, B. discolor, Cassia petersiana, Dialium schlehteri, Dictyosperma album, Diospyros mespiliformis, Encepharlatos goetzi,Englophton spp. Eugenia capensis, Faidherbia albida, Ficus sycomorus, Grewia spp. Hyphaene petersiana, Kiggelaria africana, Mimusops zeyheri, Parinari curatellifolia, Phoenix reclinata, S. tautenenii, Sclerocaryea birrea, Strychnos madagascariensis, S. cocculoides, S. spinosa, Syzygium cordatum, Tabernaemontana elegans, Tamarindus indica, Trichilia emetica, Uapaca kirkiana, Vangueria infausta, Ximenia spp. and Ziziphus species.

Non-wood Forest Products from Trees and Forests

Other than products directly obtained from forest trees, both indigenous and exotic forests are important sources of other non-wood forest products such as edible worms and insects and mushrooms. The most widely exploited worms are those of the *Imbrasia* genus (*I. belina*, *I. etali* and *I. zambeziana*) commonly found on *C. mopane*, *B. spiciformis* and *J. globiflora* and *D. mespiliformis* species. Both indigenous woodlands, particularly the miombo and the exotic pine plantations are an important source of mushrooms (Masuka (2002).

In Zambia, 38 indigenous tree species are known to produce tannins, 19 produce dyes and 11 species produce resins and gums. In South Africa, species such as*Rumohra adiantiformis* and proteas are used in the florist trade. In Namibia non-wood forest products such as beverages are estimated to have an annual economic value of N$1.5 million (US$680 000). Masuka (2002) estimated that mushroom production in Zimbabwe's pine plantations was about 807 tonnes annually although only 100 tonnes are harvested annually and exported bringing in some US$1.5 million. In Tanzania, 756 tonnes of bark of *Cinchona* spp. valued at US$258 000 was exported in 1991 (FAO, 2000). In Namibia, 600 tonnes of *Harpagophytum* spp. worth US$1.5 to 2 million in 1998 (Hailwa, 1999). In 1992, Zambia produced honey and beeswax amounting to 90 tonnes and 29 tonnes respectively which was valued at US$170 000 and US$74 000 (FAO, 2000).

Trees used for fodder

Several exotic and indigenous tree species are regarded as important fodder tree species. The indigenous tree species cited as being widely used for fodder include the African acacias (*A. erioloba, A. karroo, A. nilotica, A. robusta*) *Atriplex nummularia, Cassia petersiana, C. mopane, D. cineria, F. albida, Julbernadia paniculata, P. reclinata, Piliostigma thonningii, Swartizia madagascariensis* and *Trema orientalis*. Exotic tree species used as fodder include *A. angustissima, C. calothyrsus,Ceratonia siliqua, Chamaecytisus palmensis, L. leucocephala, L. pallida, L. diversifolia, Gliricidia sepium, Prosopsis glandiflora,* and *T. vogelli.*

Trees used for poles

Both indigenous and exotic species provide poles, which are mainly used in building huts, cattle pens, fencing and general farm construction. The most widely used species are *Androstachys johnsonii, B. plurijuga, Bivinia jalbertii, Brachystegia* spp. *C. mopane, Dalbegia melanoxylon, Dialium schlehteri, Hypaene crihata,Syzygium caudatum, Terminalia sericea*. The most durable and termite resistant species are the most preferred. Among the exotic species, the mostly widely used species for making poles are *C. cunninghamiana, E. camaldulensis, E. citriodora, E. cloeziana, E. grandis, E. nitens, E. stellulata, E. rubita* and *E. tereticornis.* In Namibia, Hailwa (1999) estimated the value of poles used for construction of kraals, hut, fencing, etc at about N$595.6 million (US$59.6 million).

Trees for shade and shelter

Both indigenous and exotic tree species are used as shade trees. The indigenous species cited as being used for shade are *A. erioloba, A. quanzensis, Albizia versicolor, Celtis africana, Olea europaea, F. albida, Halleria lucida, K. anthotheca, Kigelia Africana, S. birrea, Schinus molle* and *Trichlia emetica.* The exotic tree species used for shade and shelter are *Toona ciliata, C. cunninghamiana, Eucalyptus* spp., *J. mimosifolia, Pinus* spp. and *Thuya orientalis*

Trees used for timber

With the exception of Malawi, which has small areas of exploitable conifers such as *W. nodiflora,* most of the countries in the region do not have naturally occurring fast growing tree species that could be exploited to meet the region's growing demand for timber, pulp and tannins. The major exotic timber species are *C. lusitanica,C. tolurosa, P. elliottii, P. kesiya, P. patula, P. taeda, P. tecunumanii, P. radiata, E. tereticornis, Eucalyptus* spp and *A. mearnsii.* The commercially exploited indigenous timber species include *A. quanzensis, Allanblackia stuhlmanii, Amblygonocarpus andongensis, A. johnsonii, B. plurijuga, Balanites maughamii,Cephalosphaera usambarensis, D. melanoxylon, Diospyros* spp., *E. caudatum, Erythrophloleum suaveolens, G. coleosperma, Juniperus* spp., *K. anthotheca, Milletia stulmannii, Ocotea* spp., *Olea capensis, P. angolensis, Spirostachys africana* and *Swartzia madagascariensis.*

TREES USED FOR MEDICINAL PURPOSES

There are many tree species that are used in the treatment of a range of ailments. These species vary from country to country. The most commonly used species are*Cassipourea malosana, Cassia abbreviata, Dioscorea sylvatica, Diplorrynchus condylocarpon, Erythrophleum lasianthum, E. suavoelens, Erythrina abysinica, Harpagophyton procumbens, Ocotea bullata, Warbugia salutaris.* The value of medicinal plants in Namibia is estimated at US$3.2 million (Hailwa, 1999). In Mauritius, over 90 indigenous trees and shrubs are used for medicinal purposes. The traditional medicine industry in South Africa is estimated at between US$50 million and $100 million annually (DWAF, 1997).

Trees used for amenities and aesthetic purposes

Both indigenous and exotic species are being used as amenity trees. Among the commonly used indigenous species are *Acacia xanthophloea, Anthocleista grandiflora, Bauhinia galpinii, B. petersiana, Celtis africana, Dodonea* viscosa, *Juniperus procera, K. anthotheca, Schinus molle, Senna siamea, Trochetia boutoniana*and *Trichlia emetica.* Exotic tree species that were identified as being used as amenity trees in the SADC countries were *C. cunninghamiana, C. tolurosa, C. lusitanica, Gmelina arborea, Jacaranda mimosifolia, Pinus spp., Melia azedarach* and *Salix babylonica.* Most of these amenity tree species are either planted in gardens, along avenues in urban centres or as hedges.

Trees used for conservation/environmental protection (soil & water)

Both indigenous and exotic forest trees are used for environmental protection such as gully reclamation, restoration of degraded sites, fixing shifting dunes, etc. The commonly used indigenous tree species include African acacias (*Acacia erioloba, A. nolotica* and *A. tortillis*), *Atriplex nummularia, Burtt davyanyasica, Celtis africana, Euclea natalensis, F. albida, Hyphaene coriacea, Salix fragilis* and *Trema orientalis.* The exotic species used in environmental protection include *Acacia dealbata C. cunninghamiana* and *Populus canescens*. Some of the species like *P. canescens* are now known to be highly invasive in other countries such as South Africa and Zimbabwe were the species chokes waterways.

THE STATUS OF TREE IMPROVEMENT

The long history of selection and improvement of many agricultural crops contrasts with that for forest tree species, where the shift from exploitation to domestication is much more recent. It is generally recognised that sound management of genetic resources is critical to the optimisation of reforestation efforts - for both industrial and non-industrial applications.

For industrial forestry species, of which there are an estimated 100 million hectares of plantations worldwide, the most common approach to improvement is recurrent selection in genetically quite broad breeding populations. Four generations of breeding have been completed in *Eucalyptus grandis* and three in some *Pinus* species, but very few programmes are this advanced. The use of open-pollinated general combiner seed orchards is the most common approach to the capture of genetic gain, while a few advanced programmes are characterised by deployment of superior full-sib families or clones. Reproductive biology and patterns of variation are reasonably well documented for the established industrial species. Vigour, form and wood quality are the priority selection criteria in most programmes, with disease or insect resistance of importance in a few cases - a clear distinction from crop species, where disease and insect resistance predominate as selection criteria. The poplars are exceptional industrial species in many respects:- hybridisation and clonal selection in much narrower populations has been the traditional approach to breeding, and, perhaps not coincidentally, disease resistance as a major selection criterion. Significant genetic gains are being achieved in these breeding programmes but, in particular for the long rotation species, there has been only a minor impact to date on the genetic quality of plantations. Major limitations to rapid improvement with most of the established industrial species are:

- Long generation intervals, related to poor juvenile-mature correlations and the long juvenile phase with respect to flowering.
- Low effectiveness of selection for many characters, due to low heritability or difficulty in assessment.
- Through the use of open-pollinated orchards, the exploitation of only a part of the genetic variation available.

Major research priorities for these established industrial species should be the broader development of methods for the propagation of full-sib families and clones, and the development of methods for early and more accurate selection and the promotion of precocious flowering. While only a small proportion of current industrial plantations are located in developing countries, it is clear that much of the additional area of up to 100 million hectares required worldwide will be situated in tropical areas. Established, proven industrial species are likely to comprise a major proportion of this additional estate. A major practical tree improvement objective will be the establishment of sound breeding programmes for these new plantation operations.

In order to make use of a broader range of sites, and to supply products currently provided by harvesting in natural forests, it is likely that a significant proportion of the additional 100 million hectares will be established with tropical species which are not widely used or known at present. Selection criteria for these "new" industrial species will be similar to those for established species. Some are probably very amenable to improvement, while others present problems, e.g. flowering and seed problems, and insect susceptibility. Many potentially valuable species are simply not well known, in terms of their adaptation and biological and genetic features, and gene pools are probably under threat. The broader implementation of good tree improvement programmes will be an important priority for these species. The testing of potentially useful species, characterization of mating systems, provenance collection, establishment, maintenance and assessment of trials, implementation of gene conservation measures, and commencement of other breeding work, poses a very large task.

There is a global requirement also for several hundred million hectares at least of non-industrial plantings, mainly located in developing countries. Breeding programmes for non-industrial trees differ markedly from those for industrial species in the wide variety, and the complexity, of selection criteria. Several of the taxa are highly variable and display features which render them particularly amenable to rapid improvement by traditional means. Biological features of many potentially valuable non-industrial species, however, remain largely unknown. Although not well studied, some gene pools are under threat, and some species display unpredictable patterns of variation as a result of human disturbance. Although selection work has been undertaken in a few programmes, most non-industrial species remain at the species testing stage. The need to establish plantings on a wide range of sites, many marginal or difficult, dictates that non-industrial plantings are likely to be reliant on a substantial number of the thousands of potentially useful species. The work involved in selecting the most promising species for each site is formidable. Major emphases of improvement for non-industrial species are likely to be taxonomic studies of variation, species and provenance testing, the assessment of reproductive features, and conservation activities. Interspecific hybridization is likely to be of value for several taxa. Where regionally consistent selection criteria can be defined and the tree crop is considered sufficiently important, clonal selection programmes are likely to be of value for species easily propagated vegetatively, and seed tree selection for species propagated by seed. Sophisticated breeding programmes are unlikely to be warranted for most species. It is possible that a "once only" or "quantum leap" approach to breeding may be more appropriate than breeding programmes offering continuous incremental gains.

As pointed out by Kanowski (1993), tree improvement is greatly under-resourced. This is especially so in developing countries, where funding provided through national, regional, bilateral and international programmes is not sufficient to conduct properly the essential traditional tree improvement activities outlined above. The problems are further compounded by inadequate levels of training and facilities in most areas. For the non-industrial species in particular, it is these resource limitations and the variability in user requirements, rather than biological constraints, which constitute the major impediment to rapid improvement.

BIOTECHNOLOGIES AND THEIR APPLICATIONS TO TREE IMPROVEMENT

Cryopreservation and in Vitro Storage

For material displaying regenerative competence (e.g. embryogenic cultures), a research project directed towards the development of procedures for successful *in vitro* storage or cryopreservation of cultures for a new species might be expected to be short term (say three to five years), with a moderately high expectation of success.

Although increasingly used operationally for the storage of threatened germplasm of agricultural species, *in vitro* storage and cryopreservation have little to offer for this purpose in forest tree species. Gene pools of most of the established industrial species are reasonably well conserved in *in situ* and *ex situ* stands and preserved in seed stores. Undoubtedly, an urgent gene conservation problem exists for many tree species, particularly among the tropical hardwoods and non-industrial species. Distributions of these species are poorly known, as are their biological characteristics. Although seeds of some species will be recalcitrant, many are orthodox. Major impediments to the conservation of forest tree germplasm are the availability of resources sufficient for only a very small fraction of the survey and collection work which would be required before any germplasm could be stored, and the unreliability of many existing seed storage facilities. Replacement of existing facilities with more sophisticated technology is not likely to make a positive impact on the gene conservation problem with tropical tree species. Even for the recalcitrant species, activities should favour the establishment of *ex situ* plantings which will facilitate evaluation of the material. In the longer term, cryopreservation and *in vitro* storage may have some application as a back-up conservation strategy for populations of well surveyed species which are recalcitrant. The poplars and walnut currently fall into this category. As has been pointed out by Wang *et al.* (1993), *ex situ* conservation measures should themselves be regarded only as complementary to sound *in situ* programmes.

Cryopreservation warrants much more attention as a means of maintaining juvenility and capturing genetic gains offered by clonal forestry with industrial species. The technology is thus applicable mainly to programmes where good breeding programmes are in place, clonal forestry is a realistic goal, and "rejuvenation" is difficult - in particular for the conifers.

With the increasing emphasis on international breeding of both industrial and non-industrial forest tree species in tropical countries, international movement of material will be important. In the short term, such movement is likely to be mainly in the form of seed for species and provenance testing. Movement of vegetative material, however, is likely to become more important as breeding programmes advance, and *in vitro* approaches could find application in the longer term.

Molecular Markers

Techniques for the analysis of isozymes, RFLPs and RAPDs are now well established, and the adaptation of procedures to suit new species could be expected to be a relatively short term undertaking.

Markers have important immediate applications in supportive research for advanced breeding programmes with industrial species - mainly in relation to quality control, e.g. checking of clonal identification, orchard contamination and within-orchard mating patterns by "fingerprinting". Isozymes will be satisfactory for many, but not all, of these purposes.

Markers also have important immediate application in supportive research for tropical hardwoods and non-industrial species, in particular for essential taxonomic studies and investigations of mating systems. Both isozymes and DNA markers will be useful for these purposes. Markers will also be useful for the quantification of genetic variation to aid in sampling strategies for gene conservation and breeding population collections, although, due to modest correlations with patterns of variation for adaptive traits, they must be used conservatively as complements to traditional methods.

Realistically, application of marker-assisted selection in the short or medium term is likely to be very limited. Cheaper markers would be required and, even when these were available, the technology would apply mainly to advanced and sophisticated breeding programmes - those in which creation and maintenance of the appropriate population structures could be afforded, and where clonal forestry is achievable. A small number of programmes fall into this category and, for these, some effort aimed at developing marker-assisted selection can be justified. For most species, current resources would be far better directed towards moving breeding programmes to this stage of advancement, rather than to the development of marker assisted selection. Marker assisted selection is unlikely to have much application to non-industrial species, although, by virtue of a very short generation time, taxa such as *Gliricidia* may be useful model species for experimentation.

The major current value of markers lies in long term strategic research - in the great contributions which marker studies are making to advances in the understanding of basic genetic mechanisms and genome organization at the molecular level. For forest tree species, the study of quantitative traits will be an important emphasis of this work in coming years. This work will be most efficiently concentrated on a few model species, e.g. *Pinus taeda* and in particular hybrids such as *P. elliottii* × *P. caribaea.*

In Vitro Selection

Many recent publications concerning crop plants have reported useful correlations between *in vitro* responses and the expression of desirable field traits, most commonly disease resistance, although positive results are available also for tolerance to herbicides, metals, salt and low temperatures.

For the selection criteria of major general importance in forest tree species, in particular vigour, stem form and wood quality, however, poor correlations with field responses will limit the usefulness of *in vitro* selection. *In vitro* selection is therefore likely to have very limited application in forest tree species. It will be of possible interest in a few programmes where there is a disease resistance selection problem, but of no broad strategic value as a research objective.

Genetic Engineering

Crops transformed with genes for insect and virus resistance and resistance to various types of herbicides are at or near commercial application, and plants into which these genes have been inserted include the poplars. Many projects are in progress with forest tree species, in particular for modification of lignin biosynthesis through antisense technology. Insertion of the insect or herbicide resistance genes currently available into a new species would constitute a major research undertaking, and successful application would be dependent on being able to regenerate from the transformed cells. Manipulation of more complex traits would be a much more formidible undertaking, and, although rapid progress is being made in this field, much research remains to be done. The availability of effective transformation techniques remains an obstacle, but improved techniques are being developed. Regeneration is a difficulty for some tree species, but the problem may be over-rated - the non-competence of mature material is not necessarily an obstacle to effective application of genetic engineering, provided that juvenile material responds satisfactorily. An often overlooked research component is the testing which would be required before a responsible recommendation for large scale deployment of trangenic plants could be made. Such testing could be extensive and prolonged, depending on the species and genes involved. Research projects of this type are necessarily intensive, and must be regarded as long term with only a modest expectation of success.

Insect resistance is of potential value, e.g. in the poplars and some tropical hardwoods. A single gene may be sufficient to confer resistance for very short rotation species, but a much more cautious approach should be applied with long rotation species. The work involved in introducing several different resistance genes, sufficient to ensure that insects do not acquire tolerance during a long rotation, should not be underestimated. The reduction of lignin biosynthesis is a very valuable objective for the pulp species. The introduction of herbicide tolerance genes is of some interest, but in many programmes the advantage of using unguarded herbicide application may not be sufficient to pay for the research program. The extent to which genetic engineering permits substitution of environmentally "friendly" herbicides such as glyphosate for currently used residual herbicides could be an important factor. Cold tolerance genes are likely to be of some commercial value in many species, in particular the eucalypts. Much remains to be done, though, to establish that sufficient tolerance can be conferred using antifreeze proteins, and to extend the work to tree species. Prevention of the escape of genes into wild populations is likely to become an important concern, and sterility should be an early target of genetic engineering work with forest tree species. The major factor limiting application of genetic engineering in forest tree species is the state of knowledge of molecular control of the traits which are of most interest - those relating to growth, adaptation and stem and wood quality. Genetic engineering of these traits remains a distant prospect.

It is important that genetically engineered genotypes be of high quality with respect to other traits as well. The clonal test is the most logical basis for integration of genetic engineering into traditional tree improvement programmes. For these reasons, genetic engineering is most appropriately conducted with species where breeding programmes are advanced and clonal forestry can be realistically contemplated. Research on this subject should not assume a high priority with species for which natural variation available within the taxon remains poorly investigated.

Somaclonal Variation

Variation induced during cell or callus cultures has been reported for many species. For some crops, variants have been produced showing economically useful levels of traits such as resistance to disease and increased levels of salt. The phenomenon is still not clearly understood, and persistance of modifications through subsequent sexual generations has been demonstrated in some cases but not in others. This is not a research area where favourable results could be predicted with any confidence, and use of the approach is furthermore dependent on being able to regenerate whole plants from the cells or callus. Research with new species will be more soundly based when the phenomena are better understood in the model species already under study.

Assuming stability of the characteristics, the approach is of most value where selection can be done at the cell level (e.g. by exposure to a phytotoxin or to high mineral levels), thus enabling screening of large numbers of genotypes, and where the level of the trait sought lies outside the range available naturally in the species. Cold tolerance in the eucalypts is an example which may meet these criteria. No immediate applicability is evident to the little known tropical hardwoods or to non-industrial species, for which genetic variation naturally available is generally poorly defined.

Protoplast Fusion

This is a field with a long history of research. There have been several successes, notably with species of the Brassicaceae and Solanaceae. Severe limitations are imposed, however, by the taxonomic relationship of parents, and the requirement for regeneration from protoplasts. Expectations of success with a new pair of species could be expected to be low. This is a field with little apparent applicability to forest tree species, particularly those (e.g. some tropical hardwoods and non-industrials) for which investigations of the possibilities for traditional hybridization remain minimal. In the longer term, the objectives of protoplast fusion may be better served by manipulations at the DNA level.

Gametophyte Cultures

Although anther culture has been used for the rapid production of homozygous lines in the breeding of some self-pollinated cereals and vegetable crops, few reports exist of regeneration from gametophyte cultures of forest tree species. Research with this objective would be likely to be long-term with a relatively high risk of an unsuccessful result. Induction of haploid plants has no immediate application in forest tree improvement programmes. Such plants may be of some use in basic genetic studies, e.g. for studies of heterosis in forest tree species. As a long term strategic research objective for industrial species, induction of haploid plants should be of low priority until such time as methods for early selection and the promotion of early flowering are available. The breeding of non-industrial species is unlikely to warrant this level of sophistication.

In Vitro Embryo Rescue

This technique has been used, particularly in fruit trees, to grow embryos that normally would abort due to incompatibility between ovule and embryo development, and also for the rescue of the zygotic embryos of apomictic species. The techniques are not difficult, and development of protocols for a new species generally would be a minor research task - short term with a high expectation of success. Embryo rescue has been used occasionally in forest tree species, but the requirement for such technology is likely to be limited - to what probably will be a small number of hybrids which are sufficiently close to produce a normal embryo but for which embryo development *in vivo* is restricted. In the short term, research of this type is likely to have a very low priority. In the longer term, as barriers to natural hybridization become better defined, some work can be targeted at hybrids which species testing has identified as potentially of interest and for which other studies suggest that *in vitro* approaches might provide a solution.

Micropropagation

Over 1 000 plant species have been micropropagated, including over 100 forest tree species. It is reasonable to conclude that, with sufficient research effort, successful protocols could be developed for most tree species. In general then, a research project of this type might thus be described as short to medium term with a moderately high expectation of success.

For most industrial species, costs of planting stock and insufficient data regarding field performance remain major obstacles to be overcome before broader use of micropropagules as direct planting stock could be contemplated. Micropropagation has an immediate application in integrated clonal propagation systems featuring the commercial planting of cuttings harvested from rapidly multiplied, micropropagated stool plants of the selected clones. This approach is of value only in very advanced breeding programmes incorporating the identification of outstanding clones - currently only a few programmes with tropical eucalypts, and perhaps also some poplar programmes, but potentially also with other industrial species. Appropriate integration into breeding programmes is essential. Where clonal testing on a reasonable scale is possible and affordable, the current applicability of protocols mainly to juvenile material is not necessarily an impediment to the capture of good gains through clonal forestry. This conclusion, however, is dependent on the ability to store juvenile material for the period of a clonal test. Genetic variation in response, often substantial, is not likely to be a major problem where clonal testing can be preceded by screening for responsive genotypes, although demonstration of the absence of adverse correlations with economic traits is important. Breeding programmes with new industrial species and non-industrials are not sufficiently advanced to warrant much use of micropropagation in the short term.

Micropropagation may have wider application for the multiplication of stool plants of industrial species as breeding programmes become more advanced and other limitations to clonal forestry (e.g. maturation problems) are overcome. For some non-industrial tree species, micropropagation may ultimately have a role in the multiplication of selected varieties prior to release. Development of simple micropropagation protocols for those species for which such are not already available is therefore a useful research objective, but one which should not take priority over issues such as advancement of the breeding program.

With considerable research, developments with somatic embryogenesis and artificial seed technology may overcome the planting stock cost limitation discussed above, and enable direct use of such propagules in plantation establishment. For industrial species then, development of these technologies is a useful long term research objective, but one which is best pursued with one or two appropriate model species, e.g. *Picea abies* and *Pinus taeda.*

IN VITRO CONTROL OF THE MATURATION STATE

A long history of research with *in vitro* rejuvenation has led to some successes, but little evidence that rejuvenation can be completely, permanently and reliably achieved by this method. Similarly, sporadic reports of accelerated maturation through *in vitro* manipulations are not encouraging with respect to reliable acceleration of ageing. Further empirical work with these objectives is likely to have a low probability of success. An understanding of the molecular basis of maturation is much more likely to lead to practical manipulation, but this work is in its infancy, and acceleration or reversal of maturation to precise levels remains a distant prospect.

For clonal forestry with industrial species, maintenance of juvenility as is about as useful as rejuvenation for many purposes, and probably achievable using technologies such as cryopreservation or stimulation of juvenile stump sprouts. Nevertheless, more fundamental control of the maturation state remains one of the most valuable objectives of long-term strategic research in forest tree improvement with industrial species. Rejuvenation is most applicable to programmes where good breeding programmes are in place, and where other limitations to clonal forestry do not exist. Maturation is a much less important issue with many of the non-industrials. Manipulation of the maturation state to induce early flowering and reduce generation intervals is potentially of greater interest than rejuvenation, for industrial species at least, but only of real value where active breeding programmes are in place.

SYNTHESIS

Biotechnology represents, for many agricultural crops, the best hope for meeting the urgent objectives of breeding programmes - conservation of gene pools of wild relatives, multiplication of elite cultivars, and, in particular, the acquisition of virus and insect resistance. This applies also for many crops in developing countries, including cassava, a crop grown almost entirely by resource-poor farmers. By contrast, biotechnology offers little in relation to the most urgent priorities in forest tree improvement, although forest tree species are the subject of very active biotechnology research programmes. This is because of the major objectives of most tree improvement programmes differ greatly from those for crop species, because of the long generation times of trees, because of the great number of tree species which can provide a given end use product, because of the relatively low value of the product, and a number of other factors.

Possibilities for technology replacement or supplementation in tree improvement programmes in the short term are:

- *For breeding programmes with established industrial species:*
  - The use of molecular markers in quality control in advanced breeding programmes, e.g. for checking of clonal identification, orchard contamination and within-orchard mating patterns by "fingerprinting".
  - The use of micropropagation in integrated clonal propagation systems featuring the commercial planting of cuttings harvested from rapidly multiplied, micropropagated stool plants of the selected clones. This approach is of value only in very advanced breeding programmes incorporating the identification of high quality clones.
- *For breeding programmes with "new" industrial and non-industrial species:*
  - The use of markers in essential taxonomic studies and investigations of mating systems.
  - The use of markers for the quantification of genetic variation to aid in the design of sampling strategies for gene conservation and breeding population collections.

Strategic research priorities relating to the application of biotechnology in tree improvement are:

- *Long-term generic research. Priority objectives are:*
  - Genetic engineering for sterility. This is of high priority because it will underlie many of the eventual applications of genetic engineering.
  - The use of molecular markers and DNA transformation techniques to investigate genetic processes at the molecular level, in particular those relating to complex traits such as growth, adaptation and stem and wood quality. Of high priority, this work is of relevance in particular for industrial species, but will pave the way also for applications of biotechnology to non-industrial trees.
  - Molecular studies of the maturation state. This is of high priority for industrial plantation species.
  - Development of somatic embryogenesis, in combination with artificial seed technology, as a cheap clonal propagation method for industrial plantation species. This is of moderate priority.

Research projects of the above type are most efficiently conducted with a small number of model species. Diffusion of resources and effort to many species is likely to impede progress.

- *Long-term specific research. Some objectives are:*
  - Genetic engineering of useful traits, e.g.:
  - lignin reduction in pulp species
  - cold tolerance, particularly in eucalypts
  - insect resistance, e.g. in poplars and perhaps Meliaceae (when appropriate breeding programmes are in place)

Transformation with appropriate genes may be achieved within the short to medium term (say the next five to ten years) but must be followed by perhaps ten years of field testing before responsible commercial deployment could be recommended.

- Marker-assisted selection, for species where breeding is advanced and where creation and maintenance of the appropriate population structures is feasible and affordable. It will probably be several years before this is possible on an effective operational scale.
  - *Short- to medium-term research. Useful objectives are:*
- The examination of genetic correlations between regenerative competence and commercially important field traits. This is of high priority.
- The development of cryopreservation methods as a means of maintaining juvenility, for advanced breeding programmes with industrial species.
- The development of cryopreservation as a backup measure for gene conservation in proven species for which breeding programmes are in existence and for which seed recalcitrance has been demonstrated (moderate priority).
- The development of simple micropropagation techniques for species for which such is not already available (low to medium priority).

There are currently few areas where biotechnology can profitably replace existing technologies in tree improvement. The potential applications of several biotechnologies are interdependent e.g. the application of genetic engineering, and of cryopreservation for the retention of juvenility, are dependent on the availability of a suitable method for clonal propagation from cultured material. The applications of these technologies, and marker-assisted selection, are really only appropriate to operational, sound clonal forestry programmes, and it can be argued that clonal forestry is the gateway to major applications of known biotechnologies in forest tree improvement. Sound clonal forestry is dependent on the existence of good breeding programmes to produce genotypes for their use in well-planned and sound tree planting programmes. The funding of biotechnological research initiatives cannot be at the expense of the development of good genetic improvement programmes.

Apart from the large investments required for the development of most biotechnologies, and for appropriate testing, large investments in many tree breeding programmes are required to reach the appropriate levels of advancement, frequently with added levels of structural sophistication. As emphasized by Burdon (1992) then, a commitment to biotechnology must be part of a substantial increase in the total commitment to genetic improvement. This level of investment in breeding is not being made for most industrial species, and may never be affordable for most non-industrials. The gap between traditional methods and biotechnological approaches to forest tree breeding is therefore substantially a matter of financial investment.

The high costs of the required research programmes argue strongly for collaboration, rather than competitive proprietary biotechnology, as stressed by Burdon (1992). Certainly, it is desirable at least that expensive and long-term generic research be conducted collaboratively by specialized laboratories and with appropriate model species.

BIOTECHNOLOGY AND TREE IMPROVEMENT IN DEVELOPING COUNTRIES

Many highly profitable private forest plantation programmes exist in developing countries. Taking into account risks and potential commercial benefits, managers of these programmes will make their own decisions on the appropriate stage at which investment in biotechnology is appropriate, and the level of investment. These industrial operations aside, funding available for genetic improvement in developing countries, in particular for "new" industrial and non-industrial species, is grossly inadequate for the urgent exploration, conservation and testing work which needs to be done, and for the training and facilities which are required. Funding currently used for these purposes should be redirected to biotechnology only to the extent that effective technology replacement is achievable in the short term. In this category are the use of molecular markers in studies of reproductive biology, in taxonomic investigations, and for the quantification of genetic variation. Massive increases in funding for tree improvement in developing countries may warrant more attention to biotechnology, but diversion of currently available funds to longer term biotechnological research initiatives is likely to have a negative impact, overall, on genetic improvement programmes for these species.

APPLICATIONS IN FOREST TREE IMPROVEMENT

Three broad areas of application of cryopreservation and *in vitro* storage techniques to forest tree improvement warrant discussion:

Germplasm Preservation.

In vitro storage is now routinely used for germplasm storage of cassava, at CIAT (Escobar *et al.* 1992, Chavez *et al.* and for many *Solanum* genotypes at CIP, in Peru (Withers *et al.* 1990). ORSTOM, at Montpellier, is developing methods for the preservation of coffee by *in vitro* culture of immature zygotic embryos (Charrier *et al.* 1991). Work aimed at the storage of somatic embryos of mango is underway (IBPGR 1991). In the IBPGR *in vitro* program, high priority is being given to the development of good *in vitro* culture procedures for sweet potato, taro, cocoyam, banana and plantain, sugarcane, cocoa, citrus and forage grasses (Withers *et al.* 1990). With respect to cryopreservation, preservation of a substantial number of genotypes of oil palm by ORSTOM is the closest to routine application (Engelmann 1991), while the development of routine procedures for banana and plantain (at CATIE), cassava (at CIAT) and coconut is underway (IBPGR 1991).

A number of features characterize the above crops targeted for *in vitro* or cryopreservation approaches to gene conservation:

- They are limited in number, and are all of major, widespread importance.
- Most are characterized by a long history of domestication and use.
- Biological and agronomic features are relatively well known.
- Several are vegetatively propagated.
- Some have recalcitrant seeds.

Gene conservation in these crops is to some extent then directed towards:

- The collection of known genotypes from many different areas of use. These genotypes have been evaluated to some degree, and it is desirable gene combinations, rather than individual genes, which are the target of preservation.
- Preservation of threatened wild relatives, characteristics of which would also be partially known.

Under these circumstances, conservation priorities are reasonably easily set. For these crops, *in vitro* storage and cryopreservation offer alternatives to the traditional dependence on field genebanks. For agricultural crops where seed collection and storage are possible, however, seed storage is the recommended approach to gene conservation (Withers 1992). Undoubtedly, urgent gene conservation and preservation problems exist for forest trees species, mainly among the tropical hardwoods and non-industrial species. Compared to the crop species discussed above, the following features characterize these tree species:

- There is a large number of species of known or potential interest. Many may never be used commercially, and some will be important only locally.
- Biological features of many are poorly known.
- Species distributions, and those of relatives, are poorly known.

- Seeds of some species will be in the recalcitrant category, but many are orthodox.

Major impediments to the preservation of these gene pools are:

- Resources available are sufficient for only a very small fraction of the survey and collection work which would be required.
- Seed storage facilities in many tropical countries are unreliable, such that satisfactory storage of seed collected now cannot be guaranteed (Ng 1985, Reid *et al.* 1991).

Wang *et al.* (1993) correctly drew attention to the potential use of *in vitro* storage and cryopreservation as a complementary method for the *ex situ* conservation of long-lived perennial species. This potential is undoubtedly real. It is of value though to carry this a little further, to consider specifically to which groups of tree species the technology might be applicable in the near future. It can be observed that replacement of existing seed storage facilities with more sophisticated technology is not likely to make a positive impact on the gene conservation problem with tropical tree species. Even for the recalcitrant species, activities should favour the establishment of *ex situ* plantings which will enable evaluation of the material. Cryopreservation and *in vitro* storage techniques thus offer no solution to erosion of tropical forest gene pools. More concerted efforts at documentation and *in situ* conservation are required.

Of established industrial species, cryopreservation and *in vitro* storage of vegetative material are not advantageous approaches to gene conservation in most of the conifers and eucalypts, for which seed can be stored satisfactorily. Poplars display some of the features characteristic of the fruit tree species described above - problematic seed storage, and breeding and commercial use based substantially on known clones. *In vitro* storage or cryopreservation may have a role as a back-up measure in large poplar breeding programmes. Similarly also for other species with seed production and storage problems, e.g. *Juglans*. It should be noted though that, for most forest tree species, introduction of *in vitro* or cryopreserved material into a breeding program will involve a delay of several years to flowering, and generally also for evaluation. Flowering genotypes in field clone banks can be much more quickly incorporated. An exception to the flowering obstacle is stored pollen, which can be rapidly incorporated into breeding programmes, and pollen storage should be more prominent in gene conservation programmes with industrial species. Cryopreservation may facilitate the broader use of pollen storage in these programmes.

It should be noted also that, as pointed out by Jackson and Coleman (1991) and Wang *et al.* (1993), the use of cryopreservation and *in vitro* storage could involve selection for genotypes which tolerate the conditions imposed (e.g. freezing and thawing), and which are capable of regeneration following such treatment. The conserved material is thus not necessarily a random sample of the genetic material collected.

Maintenance of Juvenility

Cryopreservation is perhaps the most promising approach currently on offer for maintenance of the juvenile state and capture of genetic gains through clonal forestry with industrial species (Haines 1992). The technology is applicable mainly in cases where good breeding programmes are in place and clonal forestry is a realistic goal. In particular, this applies to some programmes with conifers. Although some good breeding programmes are in place for the eucalypts, the need is not so urgent for these because rejuvenation can be achieved by other means. The approach ultimately may have some application to other tropical hardwoods, as good breeding programmes are established for these. Maturation is a much less important issue with many of the non-industrials, in particular those which flower precociously, can be selected early, and which can still be readily propagated at selection age.

Transport of Germplasm

The use of plants, organs or cells cultured *in vitro* circumvents to some extent the limitations which quarantine regulations place on the import of vegetative material (in particular) into many countries. International germplasm exchanges are becoming very important in the breeding of many crops, in particular to and from international genebanks at the International Agricultural Research Centres (IARCs) and other centres. *In vitro* exchanges are becoming common for some of these, e.g. for potato, coconut, banana and oil palm (Charrier *et al.* 1991, Withers 1992). In addition, *in vitro* collection, overcoming problems of inadequate or immature seeds and deterioration of vegetative material in transit, is now being used. Simple field procedures have been devised for cocoa, avocado, coconut, citrus, cassava, cotton, and forage grasses. (Withers *et al.* 1990, Withers 1992).

With the increasing emphasis on cooperative breeding of both industrial and non-industrial forest tree species in tropical countries, international movement of material will become increasingly important. In the short term, such movement is likely to be mainly in the form of seed for species and provenance testing. Movement of vegetative material, however, may become more important as breeding programmes advance, and *in vitro* approaches could find application in the longer term.

EFFECTS OF OZONE ON FOREST TREES

FOREST RESOURCES

The forested areas within the SAMI region cover more than 50% of the land area and are very diverse in nature. Most of the tree species in the region are deciduous hardwoods with various species of oak, hickories, maples and yellow-poplar (*Liriodendron tulipifera*) as the predominant species. Many other species of hardwoods do commonly occur in the region. Although not as prevalent as hardwoods (< 30% of the total wood volume) several different coniferous species can commonly be found in the SAMI region. These include eastern hemlock, Fraser fir, red spruce, pitch pine, Virginia pine and loblolly pine. A more detailed description of the forested resources in the region is presented in *The South's Fourth Forest: Alternatives for the Future* (USDA Forest Service 1988).

Natural Stressors

Besides air pollutants, many other stresses affect forests in the SAMI region. These can be found in any forest in the world and include abiotic agents such as moisture and nutrient deficiencies (Dougherty 1995) and biotic stresses such as insects and diseases (Hepting 1971, Skelly *et al*. 1987). Commonly found insects include the balsam wooly adelgid (*Adelges piceae*), gypsy moth (*Pothetria dispar*), several bark beetles (primarily *Ips spp.*) inclusive of the southern pine beetle (*Dendroctonus frontalis*), and diseases such as sycamore anthracnose (*Gnomonia veneta*), Armillaria root rot (*Armillaria spp.*), numerous cankers, fusiform rust (*Cronartium quercuum f. sp. fusiforme*), littleleaf syndrome (*Phytophthora cinnamomi*), and annosum root rot (*Heterobasidion annosum*). Although primarily endemic in nature, insect attacks and diseases can reach epidemic proportions in the SAMI region as evidenced by the balsam wooly adelgid, gypsy moth, and the fungi which cause chestnut blight [*Cryphonectria (Endothia) parasitica*], dogwood anthracnose (*Discula destructiva*) and butternut canker (*Sirococcus clavigignenti-juglandacearum*). Wildlife activities also cause serious losses in the forest with particular concern for damage to regeneration by the whitetail deer. In addition, numerous abiotic stressors inclusive of local and/or regional scale major and minor periods of drought, rime ice and more heavy accumulations of ice leading to top breakage, freezing temperatures leading to winter injury, flooding, and of course fire are all present within the forests of the SAMI region. Moisture and nutrient stresses can be limiting factors in both the productivity and species composition of the various forest types in the SAMI region.

The interrelationships of ambient ozone exposure with all of these natural stressors remains virtually unknown. Introduced insects and pathogens usually need only a susceptible host specie(s) and favorable growing conditions for rapid speed and reproduction into epizootic/epidemic proportions. More is known about other abiotic stressors such as with drought and/or other pollutant interactions as reviewed later in this document. These naturally occurring and introduced stressors as may be found throughout the SAMI region are described in more detail by Skelly et al. (1987), Meadows and Hodges (1995) and Tainter and Baker (1996).

Objectives

The synthesis document will use information obtained from the annotated bibliography (already submitted to SAMI) to assess the scientific evidence concerning ozone effects on forest trees (primarily Class I areas) in the SAMI region. The primary emphasis of the document includes seven items: 1) evaluation of ozone effects on visible injury, growth and productivity, and physiological function; 2) review and classify historic trends, status and contribution of ozone to these trends in the SAMI region; 3) evaluate ozone exposure-response relationships derived from both greenhouse and field studies; 4) critique existing ozone exposure data sets applicably to Class I areas and their potential for use in assessing status of Class I areas; 5) evaluate other abiotic and biotic factors that may influence forest response to ozone and assess the potential influence of ozone compared with other environmental and plant factors; 6) review alternative interpretations of existing data; and 7) identify information gaps that limit assessment of current and future status of forest resources in the SAMI region. Recommendations from this assessment will be used by SAMI to address three policy questions:

- What is the current status of the resources relative to ozone effects on vegetation?
- What is the relationship between ozone exposures and vegetation response?
- What changes in resource status are projected to occur from changes in exposures due to implementation of the 1990 Clean Air Act Amendments or other emission management options being considered by SAMI's Policy Committee?

VISIBLE FOLIAR SYMPTOMS

This introductory section, as adapted in large measure from Skelly *et al.* (1987), will present descriptive information on the visible foliar injuries as induced by exposures to ambient ozone under natural field conditions within the SAMI region. It is recognized that pre-visual physiological changes are induced before the manifestation of such visible symptoms. Growth and/or productivity changes may also precede or follow visible symptoms. Please refer to the respective sections that review these forms of pre-visual and productivity changes.

Visible symptoms resulting from ozone exposure are generally recognized as resulting from either acute or chronic exposures as manifested on the foliage of sensitive plants. Injury from acute [unusually high concentrations (generally >0.10 ppm in the SAMI region), short-term (generally <24 hr)] exposures normally involves the death of cells and develops within a few hours or days following exposure. Injury is expressed as tissue bleaching, leading to bifacial necrosis (fleck), and may follow exposure to unusually high ozone concentrations. Acute ozone exposures rarely occur in nature.

Foliar injury resulting from chronic exposures [lower concentrations (generally < 0.100 ppm in the SAMI region), longer-term (generally days, weeks, or months)] may be manifested as chlorosis, pigmentation (stippling), general leaf reddening and/or premature or enhanced senescence. Both types of symptoms (resulting from acute or chronic exposures), particularly those from chronic exposures, may be confused with symptoms of other conditions, such as normal senescence, nutritional disorders, other environmental stresses, biotic pathogens, or insect infestation (Skelly *et al.* 1987).

Description of Symptoms

Broadleaf Species

Under conditions of ambient ozone exposures several specific symptoms appear on broadleaf species in the field, the most common of which is stipple, defined as discrete areas of pigmentation restricted to the adaxial leaf surface. The upper leaf surface of sensitive plants may have a tan, red, brown, purple, or black stippling (as if pepper had adhered to the upper leaf surface) that may appear uniformly over the leaf surface, or restricted to certain areas of the leaf, or discrete dot-like lesions. The lower leaf surface usually remains uninjured, with only the palisade cells underlying the upper epidermis affected. Veins and veinlets are usually not involved, and small veinlets often bound the injured areas, producing angular sections of affected tissue. Sometimes stippling is best observed by holding the leaf toward the sun; symptoms are often more intense on leaves exposed to direct light. Recent evidence suggests that shaded leaves may be more sensitive earlier in the growing season and exhibit the earliest symptoms (Fredricksen *et al.* 1995).

Stippling has been described as the classic symptom of ozone injury on broadleaf trees. The coloration of stippling is usually characteristic for a species, but can vary with environmental or physiological conditions. The youngest fully expanded leaves are normally the most sensitive, although all but the youngest leaves may be affected. On young leaves, symptoms tend to develop at the tips, on older leaves, toward the base. The entire surface of older leaves may exhibit symptoms when exposed to ozone periodically during the growing season. Although stippling is usually interveinal, following repeated and continually higher ozone exposures, extensive injury may also occur to the leaf veins themselves resulting in generalized chlorosis. In some species, especially sycamore, stippling tends to appear adjacent to the larger veins. Chlorosis, or loss of chlorophyll, may occur as a generalized condition similar to senescence, in discrete patches called mottle, or in patterns similar to stippling. Chlorosis is often more prevalent on the upper leaf surface of species that have palisade cells. On plants with pinnately compound leaves, such as ash, hickory, and tree-of-heaven, only some leaflets at certain positions along the petiole may be affected. Because stippling is a photosensitive response, overlapping leaflets or leaves may create a sharp line of demarcation between injured and uninjured (shaded) tissue.

On broadleaf species that are very sensitive to ozone, injury is usually expressed as a fleck characterized by small, discrete areas of dead cells-usually palisade parenchyma and sometimes associated epidermal tissues-leading to the formation of irregular lesions that although first evident on the adaxial leaf surface injury quickly become evident on both leaf surfaces. The lesions are often bleached (unpigmented) but can be colored as in stippling, and the affected areas may be slightly sunken. Bifacial necrosis results when the tissues connecting the upper and lower leaf surfaces are killed. The tissue coloration ranges from white to red-orange to black and is often characteristics of a specific tree species. Small veins are usually included in the necrotic tissue, but larger veins often remain alive. The area of dead cells collapses, and the upper and lower leaf surfaces adhere together to form a papery lesion.

In bifacial necrosis, the leaf margins are sometimes the most severely injured. Prolonged exposures to even low ozone concentrations may cause coalescence of chlorotic tissues, light stippling, and/or production of a bronzed appearance, especially on ozone-sensitive tree species.

Season-long exposures to ambient ozone may further result in premature or enhanced senescence of the older leaves on sensitive species. This may be a classic symptom of cumulative ozone exposures for many species, but remains elusive of description due to lack of filtered air controls for purposes of phenological comparisons. For sensitive species such as black cherry, notable stippling and general leaf reddening of the older leaves has been observed, with defoliation occurring several weeks before more tolerant individuals on the same site (Skelly, personal communication). For other sensitive individuals within species such as yellow-poplar (*Liriodendron tulipifera,* L.), white ash (*Fraxinus americana,* L.), or sassafras (*Sassafras albidum* L.) premature senescence is characterized more simply by the yellowing and dropping of the older leaves. When defoliation is severe, leaves with ozone-induced stipple may commonly be found on the forest floor below sensitive individuals several weeks before full autumn coloration.

Conifer Species

Foliar injuries on conifers as may be induced by ambient ozone exposures are far more difficult to identify in comparison to the more distinctive symptoms induced on broadleaf species. Although ozone induced symptoms were once thought to be easily diagnosed, recent controversies (Skelly, personal communication) about the causes of several needle anomalies have added to the confusion of proper diagnosis, especially under ambient forest conditions and following ambient ozone exposures. So many other causes may be involved in inducing the symptoms described below that determination of cause/effect relationships during field surveys may prove difficult.

Usually, on conifer species the two most common needle symptoms are chlorotic mottle and tipburn (Skelly *et al.* 1987). Overall, mottle of young and older needles may be induced by low ozone exposures, while tipburn of young needles may be produced by higher, early season exposures. There is, however, considerable difference in symptom expression depending on species, within-species variation, timing of exposures, and environmental conditions. Chronic symptoms usually only develop during the late summer or early fall on the current year needles with chlorotic mottle and/or spotting developing as small patches of yellow tissue interspersed with green areas (mottle), or as discrete yellow spots surrounded by apparently healthy tissue. Chronic symptoms in two, three and older-year needles simply intensify with coalescing of symptomatic areas leading to more general needle chlorosis and early senescence. Loss of second and third-year needles on eastern white pine (*Pinus strobus*, L.) and loblolly pine (*Pinus taeda*, L.) following advancing chlorotic mottle, spotting, and general chlorosis has been reported (Skelly *et al.* 1987, Chappelka and Chevone 1992).

Symptoms on most conifer species occurring as the result of acute exposures are rare and usually occur following episodic events to ambient ozone early in the growing season. If an ozone episode were to occur at this sensitive time of needle growth, that portion exposed to the higher ozone concentrations may subsequently develop necrotic needle tips resulting in tip dieback or necrotic banding. Tip necrosis is characterized with a pink to reddish coloration that fades brown with age. Towards the end of the growing season, these necrotic tips may break off, making affected needles appear much shorter. The tipburn symptom, especially in eastern white pine, usually affects all needles in a fascicle equally following an exposure to episodic ozone in the spring portion of the growing season. Severely affected individuals are evident within populations due to their off-color for much of the growing season.

On eastern white pine, a late season phenomenon of banding in the mid-portions of individual needles within the current year fascicles has also been attributed to episodic ozone exposures immediately before symptom expressions. Although shown to be induced by high ozone exposures under controlled conditions, additional controversy as to the role of fungi in causing this symptom leads to a less clear decision of determining this symptom as solely being due to ambient ozone exposures under forest conditions (Skelly, personal communication). Young, rapidly growing needles that are directly exposed to sunlight are the most sensitive, but older needles can exhibit mottle and premature senescence from prolonged, chronic exposures. Similar to broadleaf species, symptoms on plants within a single conifer species can vary considerably.

Bioindicator Species

Some particularly sensitive broadleaf tree species include black cherry, white ash, yellow poplar, and sassafras have become recognized as very useful bioindicators of ozone exposure because of a distinctive upper surface stipple, leaf reddening, and a large portion of the population appearing as sensitive under varied conditions of exposure and environments (Chappelka *et al.* 1992).

Clonal lines of eastern white pine have been suggested for use as bioindicators of ambient ozone exposures, but as described above the general use of this species under field conditions may pose difficulties in determining cause/effect relationships. Other forest species that may serve as bioindicators of ozone pollution are blackberry, milkweed, dogbane, big-leaf aster and poison-ivy. Blackberry and poison-ivy exhibit a dark purple-red stippling of the upper leaf surface that often coalesces over most of the leaf surface. The symptom on milkweed is similar to that on blackberry, except that the coloration is purple-black. Milkweed has been reported as a common bioindicator of ozone (Duchelle and Skelly 1981).

VISIBLE FOLIAR OZONE SYMPTOMS: TREE SPECIES IN THE SAMI REGION

Current knowledge of visible foliar symptoms on forest tree species growing in the SAMI region is limited to only a few species of hardwoods and conifers. Only those species that have had symptoms confirmed from laboratory studies (for example, CSTRs) or studies with open-top chambers using charcoal-filtered air controls (CF) will be discussed. For purposes of this section only open-top chamber studies and field surveys will be reviewed.

Aspen (*Populus tremuloides*): In a controlled study using open-top chambers in Switzerland, Keller (1988) observed no visible symptoms but premature defoliation occurred when ozone was greater than 20 ug/m^3-h for sensitive clones and greater than 60 ug/m^3-h for tolerant clones. In a more recent study of aspen by Karnosky *et al.* (1992), symptoms appeared as upper leaf surface stipple, bifacial necrosis, and premature leaf abscission. Foliar injury began to occur by late July and more than 50% of leaves of sensitive and intermediate clones had symptoms by the end of August. Open-top chambers were used, with cumulative ozone concentrations of 5.0, 10.0, 19.4 ppm-hr (1988) and 7.7, 15.4, 26.4 ppm-hr (1989) for 6hr/d, 3d/wk during both years. Trees were also grown in ambient air.

Black cherry (*Prunus serotina*): Black cherry develops very distinctive symptoms in the presence of elevated ambient ozone exposures. Very sensitive individuals rapidly develop an upper surface (adaxial) stipple followed by leaf reddening and early leaf senescence. In a screening study by Davis and Skelly (1992a) the predominant foliar symptom observed consisted of a dark adaxial stipple of the oldest leaves; premature senescence occurred following exposures of 12 weeks. Ozone concentrations were 0.01, 0.075, and 0.150 ppm for 6 hr/d, 2 d/wk/12 wks. Dark adaxial stipple on oldest leaves was observed with concentrations of 0.040 or 0.080 ppm ozone from 0830-1530 h / 5 d/wk for 8 or 12 wks. In a field study by Fredericksen *et al.* (1995) injury on black cherry foliage was greatest on seedlings, followed by saplings, with least injury noted on mature trees. In larger trees, symptoms were more prevalent in the lower crown.

Symptoms of red/black adaxial stipple and leaf abscission were positively correlated with ozone exposure in a study by Neufeld and Renfro (1993). Necrosis of older leaves was also observed by August. In a study in TN (Samuelson 1994a) exposed seedlings to ozone in open-top chambers with CF, 1.0, 1.5, 2.0 AA exposure treatments. A dark adaxial stippling, characteristic of ozone injury occurred on the oldest leaves of black cherry exposed to 2.0X AA from April - August 1993.

In two separate field surveys during the late summer seasons of 1991, 1992, and 1993, foliar symptoms on black cherry were observed at sites of three different elevations each in Shenandoah National Park, Va. and Great Smoky Mountains National Park, respectively. Black, red/black, and red/purple stipple was observed on black cherry foliage; symptoms were observed to have increased with increasing ozone concentration and with increasing elevation. (Chappelka *et al.* 1992, Hildebrand *et al.* 1996). Subsequent regression analysis showed symptoms to be positively correlated with SUM06 and W126 ozone concentrations, *ie.*, the higher ozone concentrations were related to the greatest amount of visible foliar injury. (Hildebrand *et al.* 1996).

Black locust (Robinia pseudoacacia): During a late season survey of several species for responses to ambient ozone exposures, significant positive correlations were shown between foliar injury (upper surface stipple), typical of ozone symptoms for this species (Duchelle et al. 1992), and elevation in field surveys in Shenandoah National Park, VA. by Winner et al. (1989).

Honeylocust (Gleditsia triacanthos): Symptoms were observed from June through September, 1981 on honeylocust seedlings grown in open plots under conditions of ambient ozone exposures in New Jersey. Symptoms included upper leaf surface stipple, chlorosis, and premature leaf drop. There was considerable variation in response among pre determined, ozone- sensitive and tolerant selections, with ozone- tolerant trees remaining green and full-crowned throughout the growing season (Smith and Brennan 1984).

Hybrid poplar (Populus maximowizii x trichocarpa): Davis and McClenahen (1993) reported that more ozone-induced injury (stippling) was observed on hybrid poplars established along ridge top than in lower elevation plots. Seedlings were planted in open plots near coal-fired power plants. More necrosis occurred in crosswind plots and injury incidence was positively correlated with the number of days when ozone was greater than 0.040 ppm. Hourly means were 0.040 - 0.050 ppm throughout the summer. There were 36 hrs with ozone concentrations greater than 0.080 ppm in June and July at one site, more than 98 hours at another site. The maximum concentration at the four sites ranged from 0.124 - 0.155 ppm.

Red maple (*Acer rubrum*): In an early study, Townsend and Dochinger (1974) reported red maple to exhibit pale green to yellowish-white chlorotic areas within the interveinal leaf tissues. Ozone fumigations were very high at 0.750 ppm for 7 hr/d, 3d/wk. There was more foliar injury in younger than in older seedlings, and more injury in seedlings from PA and MN stocks. The least injury was reported for seedlings grown from AL sources. In a later controlled CSTR study, Davis and Skelly (1992a) reported chlorosis within four weeks when wild-type PA red maple seedlings were exposed to 0.150 ppb. Ozone concentrations in all treatments were 0.0, 0.075, and 0.150 ppm for 6 hr/d, 2 d/wk for 12 wks. Samuelson (1994a) also observed ozone-induced injury on red maple as chlorosis, with more injury in the lower portions of the crown. Seedling plants were exposed to 2.0 X AA, with cumulative concentrations of 29, 38,41, 33, 32 ppm-h for April - August.

Northern red oak (*Quercus rubra*): Samuelson and Edwards (1993) reported that no visible injury was observed on mature trees or seedlings in open-top chambers with ozone concentrations of 18, 45, 87 ppm-h, or 34, 79, 147 ppm-hr and/or at 37, 95, 188 ppm-hr (0.03, 0.060, 0.0 90, 0.120 ppm for 7 hr/d/5 d/wk for 6, 8,10 wks).

Similarly, red oak was the only one of eight tree species showing no visible symptoms of ozone injury when fumigated in CSTRs at 0.010, 0.075, and 0.150 ppm 6 hr/da, 2 consecutive. da/wk for 12 wks as reported by Davis and Skelly (1992a).

Sassafras (*Sassafras albidum*): Symptoms on sassafras were observed in a field survey in Jefferson National Forest in VA. by Andersen *et. al.* (1987). Chappelka *et al.* (1992) reported increasing symptoms with increasing ozone exposure and increasing elevation in a field survey in Great Smoky Mts. National Park. Symptoms generally appeared as red to red/brown stipples on the adaxial leaf surface.

Sweetgum (*Liquidambar styraciflua*): Sweetgum exhibits a combination of symptoms including adaxial stipple and general leaf reddening. Kress and Skelly (1982) reported that adaxial black stipple increased with increased ozone concentration in CSTR chambers when ozone exposures were 0, 0.050, 0.100, 0.150 ppm for 6 hr/d for 28 consecutive days. Typical symptoms were also reported by Showman (1991) on sweetgum during a field survey conducted in Indiana in 1989. Although ozone concentrations were higher in 1988, no symptoms were noted due to the extreme drought also realized during 1988 (Showman 1991). In a controlled CSTR study, Davis and Skelly (1992a) reported a dark adaxial stipple when seedling sweetgum were exposed to 0.075 ppm after 8 wks. Plants were exposed to concentrations of 0, 0.075, 0.150 ppm for 6 hr/d for 2 consecutive days/wk for 12 wks.

Sycamore (*Platanus occidentalis*): Kress and Skelly (1982) reported adaxial stipple, and some necrosis and leaf abscission for sycamores following exposure to ozone up to 0.150 ppm for 6 hr/d for 28 consecutive days within CSTR chambers. Foliar injury increased with increasing exposures.

White ash (*Fraxinus americana*), *Green ash* (*F. pennsylvanica*): In an early investigation, Steiner and Davis (1979) observed chlorotic mottling, tan to buff colored stipple, and dark red/black stipple on white and green ash, with more injury on white ash; total leaf surface injured was 2.5 - 40%. Ozone concentrations were excessively high at 0.250 ppm for 6 hr.

In a study also conducted at excessively high ozone exposures (ozone concentrations were 0.50 ppm for 7.5 hr for 5 - 6 consecutive da, then 2 consecutive da 2 months later in CSTRs), Karnosky and Steiner (1981) reported purple or black upper leaf surface stipple, bifacial necrotic stipple, leaflet curling, and premature senescence. Total leaf surface injured was 0 - 40%. An increase in stipple as seen with increased ozone concentrations has been reported by Kress and Skelly (1982). In an open-top chamber study in New Jersey using FF, NF, and AA, 2 green ash seedlings had dark stippling late in the season (Elliott *et al.* 1987). No other symptoms were observed.

Field surveys have also demonstrated the sensitivity of these ash species to season-long ambient ozone exposures. Typical injury was seen in a field survey in Ohio in 1989. Although ozone concentrations were much higher in 1988 no significant symptoms were reported due to the extreme drought conditions which accompanied the weather systems of the 1988 vs the wetter 1989 growing season (Showman 1991). Visible injury has been observed by Bennett *et al.* (1992) in a field survey in forests of southern Indiana and southern New York in 1981, 1982, and 1986. In an intensive field survey, symptoms were observed at three sites of differing elevations in the Shenandoah National Park, VA. Symptoms increased with increasing ozone concentration and increasing elevation (Hildebrand *et al.* 1996). Symptoms were reported as a black to purple/black adaxial stipple in both 1991 and 1992.

Yellow-poplar (*Liriodendron tulipifera*): In a forest survey, Andersen *et al.* (1987) observed dark adaxial stippling of yellow-poplar foliage in the Jefferson National Forest, VA; 59% per cent of the yellow-poplar surveyed exhibited symptoms of ozone injury. Higher cumulative ozone concentrations were observed with an increase in elevation. However, Winner *et al.* (1989) did not observe increasing injury on yellow-poplar with increasing elevation in Shenandoah National Park, VA. Typical injury was observed in a field survey in Indiana in 1989. Ozone concentrations were higher in 1988, however, due to a severe drought no symptoms were observed (Showman 1991).

Davis and Skelly (1992a) reported that older leaves were more sensitive than younger leaves when exposed to ozone; ozone concentrations of 0.075 and 0.150 ppm 6 hr, 2 da/wk, for 12 wks were employed in a CSTR chamber investigation (Davis and Skelly 1992a). In a second study, dark adaxial stipple appeared on oldest leaves when exposed to ozone concentrations of 0.040 or 0.080 ppm from 0830-1530 h / 5 da/wk for 8 or 12 wks (Davis and Skelly 1992b).

In two field surveys, symptoms were observed at sites of different elevations in Shenandoah National Park, Va. and Great Smoky Mountains National Park. Symptoms increased with increasing ozone concentration and increasing elevation at GRSM but not at SHEN (Chappelka *et al.* 1992). Brown/black to brown adaxial stipple was observed on yellow-poplar foliage in 1991,

1992, and 1993 in Shenandoah National Park, Va. at three sites of differing elevations. In contrast to other studies showing increasing injury at higher elevations and ozone exposures, symptoms were most common at the low elevation and low ozone exposure site (Hildebrand *et al.*,1996). The yellow-poplar at the low elevation site were growing on a fairly wet site whereas those at the higher elevation were situated on a dry west-facing slope near the ridge top.

In contrast, one study (Endress *et al.* 1991) reported that no symptoms were induced on yellow-poplar exposed to ozone concentrations of 0.036, 0.108, or 0.172 ppm for 70 hrs with peaks of 0.085, 0.152, or 0.211 ppm.

Understory Species Sensitivities

During a vegetation survey by Anderson *et al.* (1987), ozone induced foliar symptoms were commonly seen on sassafras, blackberry, dogwood, poison ivy, milkweed, and wild grape in Jefferson National Forest, VA. Ozone symptoms has also been observed on black locust, *Clematis spp.* and *Vitis spp.* in Shenandoah National Park, Va. with increasing injury with increasing elevation and ozone exposures reported by Winner *et al.* (1989). As presented above, ozone injury was light in 1988 due to drought conditions with high ozone but in 1989 ozone induced injury was widespread on blackberry, sweetgum, yellow-poplar, wild grape (Indiana) and white ash, wild grape, and blackberry (Ohio) (Showman 1991).

When fumigated with CF, 1.0, 1.5, and 2.0X AA ozone concentrations, 25 of 28 species shown to have field injury due to ozone showed symptoms. Of 39 species fumigated, 1/3 were sensitive, 1/3 were moderately sensitive, 1/3 exhibited no symptoms (Neufeld *et al.* 1992). In a study conducted within CSTR chambers, Davis and Skelly (1992a) concluded that the observed order of ozone sensitivity from most to least sensitive tree species was observed to be black cherry, sweetgum, yellow-poplar, and white ash; symptoms were expressed on these most sensitive species as an adaxial stipple. Trees were exposed to 0.01, 0.075, 0.150 ppb ozone for 6 hr/da 2da/wk/12 wks). Within this same study, red maple and yellow birch only exhibited chlorosis while northern red oak showed no visible injury.

Conifers

Jack pine (Pinus banksiana), Eastern white pine (P. strobus): Low amounts of chlorotic mottle increased with an increasing ozone gradient on seven populations of jack pine and eastern white pine were reported following an Indiana field survey conducted by Armentano and Menges (1987). Greatest injury was seen on juvenile trees and several site factors caused other interactions. Tipburn and chlorotic mottling were seen in 23% of sampled eastern white pine stands in a field survey of 6 southeastern states. More injury occurred on plantation trees than those in natural stands, and more injury occurred on trees growing on southwest facing slopes. Highest incidence of injury occurred in Kentucky, and least in Georgia (Andersen *et al.* 1988). Due to numerous suspected causes, the foliar injury symptoms on eastern white pine have become very confusing (Skelly, personal communication). Thus, we have not included other citations at this time.

Loblolly pine (P. taeda): Meier et al. (1990) reported chlorotic mottling and flecking at 0.10 - 0.150 ppm ozone exposures within the first week, increasing in intensity and area with time. No symptoms appeared when plants were exposed to 0 or 0.050 ppm for 6 or 12 weeks. Symptoms of banded chlorosis, tipburn, and premature senescence were observed (after 3 wks in high exposures of 0.320 ppm ozone and 6 wks in 0.160 ppm ozone (Wiselogel *et al.* 1991). In a study by Shafer *et al.* (1987) open-top chambers were used and loblolly pine seedlings were exposed to AA, CF, NF, 1.25,1.5, 1.75, 2.0 X AA ozone concentrations (Shafer *et al.* 1987). Chlorotic mottle, tipburn, necrosis, and needle abscission from lower stem occurred in plots exposed to all concentrations, but were more pronounced for plants exposed to greater than ambient ozone concentrations. In a later study, Shafer and Heagle (1989) exposed loblolly pine in open-top chambers with 0.022 to 0.092 ppm seasonal mean ozone concentrations; ozone exposures were delivered at CF, NF, 1.33, 1.67, 2.0, and 3.0X AA in 1988, 1989, and 1990. Symptoms observed included chlorosis and needle necrosis which increased with increased ozone beginning at 0.06 ppm. No changes in growth occurred. Symptoms were seen in the top half of each plant. All elevated ozone treatments caused decreased foliar retention, especially after exposure to130 - 220 ppm-hr ozone. The longer seedlings stayed in CF chambers, significantly greater needle retention in 3rd growing season resulted (Stow *et al.* 1992).

Pitch pine (P. rigida): Scherzer and McClenahen (1989) observed visible injury after 4 days with fumigations of 0.30 ppm ozone. After the second fumigation at 0.20 and 0.30 ppm ozone, injury appeared as chlorotic mottle and tip burn, especially on the oldest leaves. Symptoms induced by 0.40 ppm ozone were tan flecking, mottling, and tip necrosis. McQuattie and Schier (1993) observed chlorotic tip and curvature of midsection needle portions on 20 - 25% of exposed pitch pine needles beginning with 0.10 ppm ozone exposure. No symptoms were induced at 0.050 ppm ozone, but there were 50% chlorotic needles with high exposures of 0.20 ppm ozone concentrations.

Red spruce (Picea rubens): Fincher et al. (1992), in an open-top chamber study, demonstrated that trees with winter injury had increased injury with increasing ozone concentration. Exposures were CF, NF, 2.0, 3.0, and 4.0X AA. with mean ambient ozone concentrations (NF) of 0.029 and 0.040 ppm for 24-hr and 7-hr-day, respectively. Browning of sensitive trees in spring increased with the increased ozone treatments.

No symptoms of ozone injury were observed on red spruce seedlings at Whitetop Mt, VA, when grown in exclusion chambers or in ambient air.

(Thornton *et al.* 1990). Flecking increased with needle age and on two mountains increased with elevation in a field survey in Great Smoky Mountains National Park. However, ozone was not monitored (Andersen *et al.* 1991).

General Conclusions

Tropospheric ozone air pollution has repeatedly been shown to cause foliar injury on sensitive species throughout much of the SAMI region. On broadleaf species, foliar injury has been observed as a mid- to late-season adaxial stipple, leaf reddening, and early leaf senescence. Under ambient exposures, symptoms on conifers are less evident due to many mimicking symptoms. Ozone injury has been confirmed only on a few species of trees and understory vegetation in the SAMI region. More research is need in this area.

METHODOLOGIES USED IN ASSESSING OZONE EFFECTS

Establishing the risk to forests in the SAMI region from tropospheric ozone is a difficult task. It requires approaches that utilize existing databases on the forest resources, ozone monitoring information, realistic ozone exposure-tree response experiments, modeling information to predict changes over a spatial and temporal scale, and integration of social values. All three methods have been utilized for tree species growing in the SAMI region. The first methodology involved the use of kriging (ozone concentrations) combined with data obtained from experimental results for four tree species, black cherry, northern red oak and yellow-poplar, growing in the SAMI region to determine areas of possible concern (Lefohn *et al.* 1997). Since environmental factors such as soil moisture availability may potentially influence tree response to ozone, the Palmer hydrologic index (Palmer 1965, 1967) was used to further subdivide these areas of concern.

Even though not applied, Hogsett *et al.* (1993) in the second approach discussed the use of GIS in combination with experimental data and individual and stand-level models to determine potential growth losses of several major tree species due to ozone. At risk areas could then be identified. The investigators combined exposure-response models derived from data collected in open-top chambers with predicted ozone exposures to predict growth losses. A third proposed approach used in risk assessment was developed by Luxmoore (1992). This approach is being developed for the assessment of regional responses of loblolly pine to ozone and other climatic factors (variation in rainfall) scaling up from seedlings to forest stands using various physiologically-based simulation models.

Applying Exposure Information for Identifying Areas of Concern

Lefohn *et al.* (1997) described the protocols for the identification of vegetation areas that may be at risk, which involved combining experimental exposure-response effects data for deciduous and coniferous seedlings and trees with 1) kriged characterized O_3 ambient exposure data mentioned previously and 2) soil moisture. The conclusions of Musselman *et al.* (1994) and others (see U.S. EPA 1996), that all hourly average concentrations have the potential for impacting vegetation, but that the higher values should be given a greater weighting than the mid- and low-levels, were used in the methodology described by Lefohn *et al.* (1997). Recent work by Karnosky *et al.* (1996) appears to reinforce the importance of high hourly average concentrations. The cumulative-type exposure index performs adequately in relating growth reduction to vegetation and ozone exposures occurring with single experiments (U.S. EPA 1992, 1996, Lee *et al.* 1991, Lefohn 1992). However, when attempting to relate a particular set of exposure-response results to ambient conditions or other experimental results, single-parameter cumulative indices should be combined with some measure of the high hourly average values (e.g., values ^{3} 0.10 ppm), which occurred in many of the open-top experiments (Lefohn and Foley 1992, 1993, Lefohn *et al.* 1992b, Lefohn 1992).

In their analysis, the 24-h sigmoidally weighted exposure index (Lefohn and Runeckles 1987), W126, was used over the April-October period for the reasons described above. Alternatively, a 24-h SUM06 (the sum of all hourly average concentrations ^{3} 0.06 ppm) exposure index could have been used. Both the W126 and the SUM06 are highly correlated and provide similar exposure-response results in effects modeling efforts (U.S. EPA 1996). The W126 was selected, however, because it does not use a subjectively determined threshold of 0.06 ppm, which cannot be biologically substantiated at this time. The index includes the lower, less biologically effective concentrations. The W126 cumulative index was integrated over a 24-h period.

For estimating the exposure regimes that relate to growth reduction of various deciduous and coniferous tree species grown in the Southern Appalachian region, Lefohn *et al.* (1997) characterized ozone exposures from biological experiments. The initial review of the literature associated with this research effort focused on exposure-response information for the following species: black cherry; yellow-poplar, Virginia pine; red maple, eastern white pine, slash pine, pitch pine, loblolly pine, northern red oak, white oak, sycamore, shagbark hickory, American beech, white ash, and green ash.

Using the experimental results, four broad sensitivity categories were developed [*i.e.*, 1) minimal, 2) level 1 (black cherry), 3) level 2 (yellow-poplar), and 4) level 3 (northern red oak, Virginia pine, loblolly pine, eastern white pine, white oak, sugar maple, red maple, American beech, and shagbark hickory] were defined. For assigning the response categories, focus was on only those studies where growth effects were noted. The minimal category was defined as a level where exposures were less than those which cause damage to black cherry or where moisture conditions were not favorable for ozone uptake.

The initial approach assumed that the 1) environmental conditions were favorable for ozone to enter the leaf and 2) total cumulative exposure would result in a growth loss. However, as mentioned above, it is necessary to consider conditions that affect a plant's sensitivity. The experimental studies used seedling and tree plants which were grown under optimum conditions (e.g., adequate moisture and nutrients). Showman (1991), Jackson *et al.* (1992) and Kouterick (1995) have observed significantly fewer ozone symptoms on sensitive species during periods of drought than during years when the growing season had adequate rainfall. Unfortunately, at this time, little experimental information is available relating ozone exposure, drought conditions, and tree growth reduction.

Each ½° by ½° grid cell in the Southern Appalachian area was assigned one of four categories (Table 3) mentioned above. Because the criteria listed in Table 3 requires that both the W126 and number of hours ^{3} 0.10 ppm be met, it was necessary to predict the number of occurrences of high hourly average concentrations. There is a paucity of air quality monitoring data which makes it difficult, at this time, to spatially predict the number of hourly average concentrations ^{3} 0.10 ppm accurately. However, the authors found that it was possible to separate the area into broad exposure categories due to the occurrences of hourly average concentrations ^{3} 0.10 ppm during "high" and "low" ozone exposure years.

For example, in 1983-1986 and 1989-1990, the number of hourly average concentrations ^{3} 0.10 ppm at all sites in the geographic area was less than 51, which is below the Level 2 sensitivity category. In 1987, there was only one site that experienced greater than 51 occurrences ^{3} 0.10 ppm. In 1988, the high exposure year, 11 of 15 monitoring sites experienced 51 or more hourly occurrences ^{3} 0.10 ppm. Subjectively, it was decided that grids which had two or more ozone monitors were classified using the highest value; cells which did not have a monitor were classified by examining the pattern from monitors surrounding the cell and selecting a site whose value was the second highest number of hours ^{3} 0.10 ppm. The Palmer hydrologic index was selected as an indicator of soil moisture (Palmer 1965, 1967). Because Palmer hydrologic index data were available for the period 1983-1990, the study was limited to this time frame. The index is a monthly value, computed for a climatic division, which indicates the severity of a wet or dry spell.

Combining the Palmer hydrologic index and ozone exposures allowed Lefohn and co-workers (1996) to identify those areas within the region where 1) soil moisture may have been adequate in the area and 2) ambient exposure regimes closely matched those experiments where growth losses were observed. Areas which were classified as experiencing a drought were assigned the minimal category; otherwise the sensitivity category value remained the same after applying the criteria previously mentioned. Using available monitoring data for 1983-1990 and exposure-response data based on seedlings and trees, Lefohn *et al.* (1997) identified geographic regions within the area that may have experienced ozone exposures, which include high cumulative values as well as the presence of sufficient numbers of high concentrations, coupled with sufficient soil moisture, that have the potential for inhibiting vegetation growth. As indicated earlier, current ozone exposures are causing visible symptoms on the foliage of sensitive species and injury has been documented in numerous locations throughout the Southern Appalachian area.

Their results indicated that in a small number of areas within the region, ozone exposures and soil moisture availability might be sufficient to cause growth losses to some sensitive species. The number of hectares where vegetation may be affected by exposures may represent an overestimate due to the optimum growth conditions experienced in the experimental open-top chambers and the manner in which the investigators characterized the ambient ozone data (i.e., over a 7-month period). In addition, two other items are important: 1) the growing range of each species and amounts of species in each cell were not used in our analysis; and 2) the resolution of the Palmer hydrologic index is at the climatic division for each month and, depending upon the variability of soils in the climatic area, the index may provide less than optimum predictions. It is clear that there still needs to be verification of actual growth losses for the areas of concern.

Risk Characterization

To assess potential risk of sensitive tree species to ozone, Hogsett *et al.* (1993) applied an assessment methodology based on the use of a Geographical Information System (GIS) combined with estimated ozone exposures, simulation models and experimental data. To test this method, they used the eastern US, including the SAMI region as the study area. For the purposes of this initial, preliminary assessment eight tree species were used: quaking aspen, black cherry, yellow-poplar, sugar maple, red maple, loblolly pine, Virginia pine, and eastern white pine. All species occur in the SAMI region, however, quaking aspen and loblolly pine are minor components of the forest types commonly found in this region.

To characterize risk of these forests to ozone, Hogsett *et al.* (1993) attempted to integrate four elements to form the data base that comprised the two major components of the assessment (exposure characterization and characterization of ecological effects). These four elements are:

1. Ozone exposures for the region.
2. Spatial distribution of the resources. The degree of interest can be at the species, community or ecosystem level.
3. The spatial distribution of other abiotic or biotic factors that may influence tree response to ozone. Included in this is environmental factors and exposure dynamics.

4. Exposure-response functions for the selected measurement endpoints.

The framework of this assessment is illustrated in Figure 11. (Hogsett *et al.* 1993). Information used to develop the response functions (ecological effects) are enclosed within the circles. These data are derived through experimental data or literature, then simulated over time by using simulation, process-level models. Models were used because all of the data used in the preliminary assessment was from individual tree seedling data gathered over a short duration (generally one year or less). These results (response functions) were then integrated with ozone exposure information, species distributions and other environmental factors using GIS to derive the final risk characterization, ie., the likelihood of the occurrence of adverse effects (biomass reductions, etc.) with ozone exposure (Figure 11).

Since ozone monitoring data are limited, ozone exposures were estimated for areas without monitors using a method developed by Hogsett and Herstrom (1991, Hogsett *et al.* 1993). The method uses GIS to generate potential ozone exposures based on ozone precursors (NO_x), solar radiation, stagnating air masses and wind direction. The ozone exposure statistic used was the maximum 3-month SUM06 value. Exposure-response information was developed from open-top fumigation experiments. The results of this preliminary assessment (Hogsett *et al.* 1993) indicated a wide range of annual biomass losses (0-33%) depending on species and ozone exposures. The two most sensitive species were black cherry and aspen with annual predicted growth losses over 20% for greater than 50% of their range. Yellow-poplar, loblolly pine, sugar maple and white pine had between 5-12% annual biomass losses with a large portion or their range experiencing these losses.

The authors (Hogsett *et al.* 1993) this assessment was preliminary in nature and subject to several sources of error. These include uncertainty associated with the use of seedling data to estimate mature forest response, climate data, estimation of ozone exposures, and species distribution. The authors do feel, however, that this is a valid technique and deserves further refinement. One key difference between that by Hogsett *et al.* (1993) and Lefohn *et al.* (1997) is that Hogsett *et al.* (1993) did not focus on the presence of high hourly average concentrations (i.e., ^{3} 0.10 ppm), while Lefohn *et al.* (1996) did. As discussed in Lefohn *et al.* (1997), many of the experiments used to develop the exposure-response information by Hogsett *et al.* (1993) were based on NCLAN-type exposures, which experience numerous occurrences ^{3} 0.10 ppm in the experimental treatments where effects were observed. Ignoring the presence of the high hourly average concentrations, may lead to overestimates of growth reduction. In addition, Hogsett *et al.* (1993) used a 92-day SUM06 exposure index. The rationale for using a this exposure period is unclear. The growing season for most tree species is generally greater than this three-month period.

Scaling Up Physiological Responses to Ozone

Another potential assessment methodology to consider is one proposed by Luxmoore (1992) where the response of loblolly pine variations in rainfall and ozone exposures would be scaled-up from a seedling to a forest stand and long-term effects on loblolly pine productivity would be assessed. The framework for this proposed methodology is shown. Input variables are first entered into a physiologically-based model [Unified Transport Model (UTM)]. This model consists of five linked models that provide data at an hourly basis on carbon, nutrients, water and pollutant fluxes to trees in a uniform stand. Data are then incorporated in the FORET model (previously described) and trees grown over time. This in turn is linked to PTAEDA2 to provide different management options and stand manipulations. These are then incorporated with different policy options with different pollution and climate change scenarios (Figure 12).

This assessment technique is still in the development stages, *ie.,* it has been tested on only one tree species (loblolly pine). It, however, allows for the incorporation of policy options to a physiologically-based process model. The model allows for the manipulation of various edaphic and climatic variables, as well as ozone and carbon dioxide levels.

7

In Vitro Propagation of Forestry

CRYOPRESERVATION AND IN VITRO STORAGE

PRINCIPLES AND EXPERIMENTAL ACHIEVEMENTS

In Vitro Storage

As detailed elsewhere in this report, micropropagation procedures have been developed for over 1 000 plant species, many of which are micropropagated commercially. Rapid multiplication, involving rapid growth and frequent subculture, generally is an objective of commercial micropropagation. The basis of successful storage of cultures is an increase in the interval between subcultures by reduction of growth. Methods employed to this end have been well reviewed by Engelmann (1991), Withers (1992) and Wang *et al.* (1993)[1]. Alternatives are:

- Temperature reduction. This is the most common approach to storage of cultures. A limitation for tropical species is that many are intolerant of low temperatures. Culture temperature must be over 20°C for cassava, and over 18°C for oil palm (Engelmann 1991).
- Reduction of light intensity.
- Alterations to the culture medium. Such alterations include reductions in the sugar and/or mineral composition, the addition of cryoprotective or osmotic agents such as mannitol, the addition of growth retardants such as abscisic acid, and the addition of activated charcoal. Additions to the medium have to be tested carefully - osmotic agents can have detrimental effects on the tissues of some species (Engelmann 1991), and cultures of tropical species frequently respond poorly to the addition of abscisic acid (Charrier *et al.* 1991).
- Modification of the gaseous environment, in particular the lowering of oxygen content (frequently achieved by covering the culture with a thin layer of oil).

Storage of cultures has involved explants of various types. Shoots with roots have demonstrated better survival than those without, and microtubers are suited to storage for potato (Engelmann 1991). An innovation which holds promise for some species is the encapsulation of explants in alginate beads. In work reviewed by Engelmann (1991), encapsulated mulberry buds and sandalwood embryos survived for 45 days at 4 degrees, and resumed growth after storage.

Withers (1992) tabulates some 30 species for which successful slow growth storage, from several months to several years, has been reported. Success has been achieved with several woody species, including *Pinus radiata, Alnus glutinosa,* and species of *Eucalyptus* and *Populus*. Research aimed at developing *in vitro* storage protocols is largely empirical, involving the testing of various treatments and environmental conditions. It is likely though that, with sufficient effort, *in vitro* storage procedures could be developed for many of the tree species for which micropropagation protocols exist. Some tropical species may be difficult, however, to store in this way. *In vitro* storage is now routinely used for germplasm storage of some crops. The *in vitro* cassava gene bank at CIAT, Colombia, comprises nearly 5 000 clones, in an area of 50 square metres, with transfer (subculture) intervals of 12–14 months (Escobar *et al.* 1992). While *in vitro* storage thus offers some advantages over field genebanks, the management of large collections remains problematical, due to the requirement for periodic subculture. The possible introduction of genetic variants during culture may be a risk with some types of cultures (Withers 1992).

Cryopreservation

Although some microorganisms and higher plant pollen can be freeze dried, suspension of growth in higher plant somatic tissues can be achieved only by transfer to ultra low temperatures such as that of liquid nitrogen (Withers *et al.* 1990, Wang *et al.* 1993). At these temperatures, metabolism is suspended, and free radical damage caused by ionizing background radiation is the only factor potentially causing deterioration (Withers 1992). Results indicate that exposure to such temperatures need not be intrinsically damaging, although transitions must be carefully managed. Cryopreservation procedures thus involve the successive steps of choice of material, pretreatment, freezing, storage, thawing, and post-treatment handling, for which optima must be defined for each species (Engelmann 1991).

Growth recovery in cryopreserved cell suspensions takes place if a certain percentage of cells survives (Charrier *et al.* 1991, Dereuddre 1992), and reports of successful cryopreservation most commonly concern cell suspensions. Maintenance of structural integrity is more important with organized tissues such as meristem and somatic, pollinic or zygotic embryos, and reports of successful cryopreservation of these are less common. Meristematic cells are the most resilient, and young somatic embryos or immature zygotic embryos are preferred (Engelmann 1992). In oil palm, somatic embryos of only a specific morphological type can be cryopreserved (Engelmann 1991).

Pretreatment comprise the application of a cryoprotectant, dehydration, or a combination of both. Pretreatment with cryoprotectants such as sucrose, sorbitol, mannitol, dimethylsulfoxide, or polyethylene glycol dehydrates tissues, but may also act by protecting membranes and enzymatic binding sites from injury. The appropriate type, concentration and duration of treatment varies with species (Engelmann 1991). The optimal duration of desiccation treatment varies from two to four hours, during which water content is reduced to 10–16% (Engelmann 1992).

A two stage cooling process is most commonly employed. Cultures are cooled gradually down to a specific temperature, and then immersed in liquid nitrogen. Species vary in their requirements with respect to both the rate at which this cooling is conducted, and the temperature to be reached, e.g. *Citrus* embryos have to be cooled at a very specific rate down to -42°C, while those of oil palm can withstand a wide range of freezing rates (Engelmann 1992). With this approach, ice formation occurs first in the external medium, and water flows out of the cells. Cells reach a level of dehydration at which crystallization of the remaining water causes no damage (Engelmann 1991). Programmable freezing apparatus is required for this work. For certain tissues of some species, a one step rapid freezing, accomplished by direct immersion in liquid nitrogen, is satisfactory. In this case, intracellular ice crystallizes as microcrystals which are not harmful to cell components (Engelmann 1991). A third alternative is "vitrification", whereby water forms an amorphous glassy structure, without a crystallization phase. This is achieved by the use of very high concentrations of cryoprotectants and very rapid cooling. This process has been applied successfully to a range of cell suspension, protoplast and embryogenic cultures, and also to shoot cultures of asparagus, mint, orange, carnation, potato and white clover (Dereuddre 1992, Withers 1992).

Thawing after storage is generally rapid, to avoid recrystallization. After the freeze-thaw cycle, culture conditions different from those of the standard procedure may have to be defined in order to stimulate the regrowth of the material. A recent innovation has been the incorporation of alginate bead encapsulation methods into cryopreservation procedures (Dereuddre 1992). This method offers protection from adverse effects of some of the treatments. Somatic embryos or shoot tips are suspended in culture medium supplemented with 3% Na-alginate. The mixture is then dispensed from a pipette. Beads of about 4 mm diameter containing 1 to 3 embryos or shoot tips are precultured for several days in media supplemented with sucrose. Coated organs are dried under sterile air flow at ambient temperature and humidity for up to six hours. After dehydration, artificial seeds are transferred into cryovials without liquid medium for cooling, using either a one or two step procedure. The encapsulation-dehydration procedure has been applied to cryopreservation of somatic embryos of carrot and shoot tips of pear, potato, grape and carnation.

Cryopreservation has been applied successfully to over 70 plant species (Engelmann 1992, Withers 1992), including tropical crops such as coconut, rubber, cocoa, oil palm, peanut, rice, banana, cassava, and coffee. Forest tree material for which successful cryopreservation has been reported includes embryos of *Quercus petraea, Fagus sylvatica, Aesculus hippocastanum*(Jorgensen 1990), *Araucaria excelsa, Castanea* (Engelmann 1992), *Artocarpus, and Juglans* (Withers 1992); embryogenic cell lines of *Picea glauca, Acer pseudoplatanus* (Withers 1992),*Picea abies* (Durzan 1988, Bercetche *et al.* 1990, Gupta *et al.* 1987) and *Pinus taeda* (Gupta *et al.* 1987, Durzan 1988), seeds of *Abies alba, Sequoiadendron giganteum, Larix decidua, Pseudotsuga menziesii, Picea abies, Pinus sylvestris* (but not *Quercus petraea*) (Jorgensen 1990), pollen of *Betula pendula, B.pubescens, Larix decidua, L.kaempferi, Pseudotsuga menziesii, Picea abies, Pinus sylvestris, Quercus petraea* and *Q. robur* (Jorgensen 1990), and encapsulated shoot tips of *Eucalyptus gunnii* (Monod *et al.* 1992). The expectation is that, once cryopreserved, material can be stored indefinitely in liquid nitrogen without further loss. Strawberry and peanut meristem were capable of regeneration after two years storage, and 1.5 years for oil palm, while potato and cassava meristem survived for four years (Engelmann 1991). Cell lines of *Picea glauca* have shown no loss of viability after over two years of storage (Attree & Fowke 1991).

The main emphasis of most cryopreservation research programmes lies in defining non-damaging procedures for cooling and rewarming the cultures, for which species have differing requirements. Rapid progress is being made with this work, and it seems likely that, given a sufficiently intensive research effort, prospects for developing a suitable protocol for tissue preservation for any chosen species within the short term would be reasonably good. One obstacle to application for many species lies in the need to regenerate plants from the tissues or organs after the period of cryopreservation.

IN VITRO SELECTION

PRINCIPLES AND EXPERIMENTAL ACHIEVEMENTS

This concerns the identification, using *in vitro* selection procedures, of desirable genotypes with respect to field performance traits such as growth, production of useful metabolites, and resistance to stresses caused by agents such as salt, heat, cold, drought, disease, insects, metals and herbicides. Selection *in vitro* can be at the level of cells (or protoplasts), microspores, buds, shoots, embryos, or whole plants. Selection at the cell level is attractive in terms of the large numbers that can be screened, but involves the particular problems of the generation of genetic variation, and sometimes the requirement to regenerate from the selected cells. Selection procedures reviewed here are considered independently of the source of variation, although a majority of reports concern putative genetic variation induced *in vitro* directly.

Selection for Disease Resistance

The various approaches to *in vitro* selection for disease resistance have been reviewed by van den Bulk (1991). Cultures can be exposed to a toxin, toxin analogues, filtrate or the pathogen itself. The use of purified toxins as selection agents in culture is potentially effective when symptoms of the disease are caused by a toxin produced by the pathogen, and where the toxin operates at the level of the explant cultured (e.g. at the cell level). Selection with crude filtrates has been used in many studies, in particular when a filtrate exhibits phytotoxic activity but no well characterized toxin is known. Selection by co-culture with the pathogen has been of limited success, in particular due to difficulties in growing or controlling the growth of the pathogen in culture. Corn plants regenerated from callus lines selected for resistance to the purified toxin from *Helminthosporium maydis* displayed resistance to the pathogen (Widholm 1988). Resistant plants were also obtained using selection on the basis of the insecticide methomyl, apparently acting as a toxin analogue (Kuehnle 1990, Kuehnle & Earle 1992). Elm microcuttings displayed wilting in response to both filtrate and coculture with *Ophiostoma ulmi* (Dorion & Bigot 1987, Dorion *et al.* 1988).

In another experiment, rate of fungal growth in coculture with callus from mature trees was correlated with known field resistance of the trees to Dutch elm disease (Dormir 1992). Resistance of cultured loblolly pine embryos to infection by *Cronartium quercuum* spores showed high family correlations with field resistance (Frampton *et al.* 1983). The ranking of poplar clones for field resistance to *Septoria musiva* was similar to that derived by inoculating cultured leaf disks with spores (Ostry *et al.* 1988). A high correlation between callus resistance to filtrate and resistance of regenerated plants to *Fusarium oxysporum* was reported for alfalfa (Arcioni *et al.*1987, McCoy 1988). *In vitro* resistance of potato plantlets cocultured with *Phytophthora infestans* was well correlated with known field resistance of cultivars (Tegera & Meulemans 1985). Protoplasts of two grape varieties reputedly resistant to *Eutypa lata* showed good viability in the presence of the filtrate, while those of three reputedly sensitive varieties died rapidly (Mauro *et al.* 1986). Rankings determined with *in vitro* inoculation of leaf disks reflected well field resistance of poplar clones to rust (Singh & Heather 1982).

Tomato plants regenerated from callus tissues resistant to toxins secreted into media displayed resistance to *Pseudomonas solanacearum* (Toyoda *et al.* 1989). Resistance of callus tissues to filtrates from *P. syringae* has been used as an assay for bean cultivars resistant to halo blight (Hartmann *et al.* 1986).

Insensitivity of protoplasts to methionine sulfoxymine, a toxin analogue, has been used to regenerate tobacco plants resistant to wildfire disease caused by *P. syringae*. Sensitivity of poplar microcuttings cocultured with *Xanthomonas populi* was well correlated with clonal susceptibility to bacterial canker in field tests (Janssen 1989). Toxin resistant cells selected in callus cultures of a peach clone highly susceptible to bacterial leaf spot disease yielded plants some of which displayed much higher field resistance (Hammerschlag 1986). Direct inoculation of pear plantlets with *Erwinia amylovora* provided a resistance assay well correlated with field resistance to fire blight (Viseur *et al.* 1987). Selection of cell lines resistant to culture filtrate yielded regenerated plants with increased resistance to *Phoma lingam*(Sacristan 1985).

As indicated, many of the above involve selection among populations of somaclonal variant cells. There are many reports also of cell lines selected by exposure to toxin yielding regenerants which do not display increased resistance to the disease. Although possible instability of the variants is a confounding factor in many of these studies, some undoubtedly result from the fact that some diseases do not act at the cell level.

Selection for Herbicide Tolerance

In principle, this method is straightforward. Herbicide is simply included in the medium at an appropriate concentration. Selection operates on the physiological basis that many herbicides also directly kill cells and tissues. Some herbicides, however, act by interfering with the functions of organised tissues, e.g. photosynthesis, and may not affect the growth of heterotrophic cell cultures (Chaleff 1986, Widholm 1988). Similarly, anatomical features such as the cuticle may influence *in vivo* but not *in vitro* absorption of herbicide (Smeda & Weller 1991). Most available published reports concern the selection of somaclonal variant cell lines showing resistance to herbicide. An exception is that of Yenne *et al.* (1987), where responses of existing commercial cultivars were compared. In another interesting exception, rape microspores were subjected to mutagenic agents, plated, and then early stage embryos subjected to range of herbicides. Survivors completed the regeneration and colchicine doubling phase and plants were regenerated for testing (Beversdorf & Kott 1987)

Resistance to most major herbicide classes has been selected for *in vitro*, and at least partially resistant plants regenerated in many cases. Atrazine-resistant plants were regenerated from cells selected in green cultures of *Nicotiana plumbaginifolia* (Cseplo *et al.* 1985). In another study, cotyledonary node plus epicotyl explants of soybean were cultured on atrazine-containing medium. Some explants yielded organogenic shoots from which plants were regenerated. Some of these plants displayed resistance to atrazine (Wrather & Freytag 1991). Calli of *Nicotiana debneyi* displaying very high levels of tolerance to amitrole were selected by stepwise exposure to increasing concentrations. Regenerated plants displayed tolerance as calli (Swartzberg *et al.*1985). Amitrole tolerance at the whole plant level was reported in tobacco plants regenerated from cell lines selected *in vitro* (Chaleff 1986).

A similar result was achieved with maize (Anderson*et al.* 1987). Plants regenerated from resistant cell lines of tobacco displayed tolerance to the sulfonylurea herbicides chlorsulfuron and sulfometuron methyl (Chaleff 1986, 1986b). Plants regenerated from hybrid poplar leaf explants subjected to selection on media containing sulfometuron methyl were tolerant of herbicide levels lethal to control plants (Michler & Haissig 1988, Michler 1988). Regeneration of shoots from poplar leaf explants exposed to glyphosate gave rise to glyphosate tolerant plants (Michler & Haissig 1988). In pea, *in vitro* sensitivity of some commercial cultivars showed some correlation with field sensitivity to glyphosate (Yenne *et al.* 1987). Tobacco plants tolerant to picloram were regenerated from resistant cell lines (Chaleff 1986). For tomato, some tolerance to paraquat was recorded in plants regenerated from cells selected on media containing this herbicide (Chaleff 1986).

Instability of the tolerance is a feature of several of these reports, and there are several examples of poor correlations. In birdsfoot trefoil (*Lotus*), for example, plants regenerated from cells surviving in chlorsulfur on media actually were more sensitive than the controls (MacLean & Grant 1987). There are many examples also of resistant cell lines whose capacity to give rise to resistant plants has yet to be tested.

Selection for Salt Tolerance

This is also a simple procedure — plants are regenerated from explants displaying tolerance to salt added to tissue culture media. Salt resistance in plants includes both avoidance and tolerance mechanisms. Resistance in *Atriplex* for example depends on the anatomical and physiological integrity of the whole plant and not on cellular properties. Nevertheless, cellular tolerance mechanisms exist which are more amenable to *in vitro* selection techniques. Elevated levels of the amino acid proline are believed to protect plant tissues against stress by acting as N-storage compound, osmosolute and hydrophobic protectant for enzymes and cellular structures, and proline effected salt tolerance has been reported in several crops (Jain *et al.* 1991). Cell lines tolerant of elevated levels of salt in the medium have been selected in many studies. In some of these studies, plants regenerated from the cell lines have also displayed increased tolerance in greenhouse and field trials, e.g.: alfalfa (Winicov 1990, 1991), *Coleus blumei* (Ibrahim *et al.* 1992), *Brassica juncea* (Jain *et al.* 1991) and *Citrus sinensis* (Spiegel-Roy & Thorpe 1986). Plants regenerated did not display such tolerance in tobacco (Watad *et al.* 1991). All of the these reports concern variation induced in culture. Stability of tolerance through subsequent sexual generations has been evident in some studies (Ibrahim *et al.* 1992, Jain *et al.* 1991), but not in others (Lucas *et al.* 1989).

Selection for Tolerance to Metals

Once again, this is simple in principle - explants are cultured on media containing high levels of metal salts. In practice, however, complex interactions among various nutrients, and pH effects, must be taken into account. As for salt tolerance, plants can resist high levels of metals such as aluminium and cadmium by various avoidance (e.g. decreased uptake) or tolerance (e.g. altered enzyme structures, precipitation in cytoplasm) mechanisms (Tal 1983). These operate at different levels, and therefore good correlations between *in vitro* and whole plant responses are not necessarily expected. Cell lines resistant to elevated concentrations of metal ions have been selected in many studies, including those involving aluminium, cadmium, mercury, zinc, lead, and manganese. Increased tolerance to aluminium in subsequently regenerated plants was demonstrated for *Nicotiana plumbaginifolia* (Meredith *et al.* 1988) and rice (Van Sint Jan & Bouharmont 1992). Most studies involve selection among variants induced by the culture process. Meredith *et al.* (1988), however, cited some studies where responses of callus cultured from cultivars known to vary in sensitivity to metals were examined. Differences among *Agrostis* genotypes in resistance to zinc were maintained in callus cultures (Wu & Antonovics 1978), while cultivar differences in sensitivity to zinc (Christianson 1979), and manganese (Petolino and Collins 1985), were not correlated with callus responses for bean and tobacco respectively.

Selection for Tolerance to High Temperatures

Cultures are exposed to high temperatures and survivors selected. Tolerance is presumably based on reduced incidence of or resistance to the effects of protein denaturation, reported to be the primary cause of heat injury (Quamme & Stushnoff 1983). Plants regenerated from cotton cell cultures exposed to regular high temperature treatments themselves yielded callus with an elevated tolerance to high temperatures (Trolinder & Shang 1991), although whole plant responses were apparently not examined in this study.

Selection for Tolerance to Low Temperatures

For this purpose, explants are exposed to low temperatures and survivors selected. Methods include those involving exposure to the frost temperatures expected, through to immersion in liquid nitrogen. Plants are unable to avoid the low temperatures, and only tolerance mechanisms enable the plant to survive adverse effects. Resistance operates mainly at the cellular level, e.g. accumulation of antifreeze substances, dehydration to minimize freezable water and increased capacity for supercooling, and good correlations between cellular and whole plant responses are therefore expected (Tal 1983). The involvement of proline has been reported (Teulieres *et al.* 1989, Tantau & Dorffling 1991). Chilling tolerance is apparently an additive, multigenic character, and different mechanisms control resistance at different developmental stages (Paull *et al.* 1979). For eucalypt clones representing a range of taxa, viability of protoplasts after cooling to -10°C was well correlated with field resistance determined by assessing sprout regrowth after frost injury (Teulieres *et al.* 1989). Nowak *et al.* (1992) concluded that testing the ability of cultured whole plantlets to withstand freezing could be used to rank alfalfa germplasm for cold tolerance. Some callus lines surviving immersion in liquid nitrogen gave rise to plants significantly more tolerant of low temperatures than unselected controls (Kendall 1991).

Selection for Tolerance to Water Stress

Methods employed have generally involved the attempted selection of cell lines showing tolerance to osmotic agents such as polyethylene glycol. Drought resistance in plants can operate through avoidance and tolerance mechanisms. Avoidance involves the maintenance of high internal water potential in the presence of external water stress, while drought tolerators are able to endure dehydration of the protoplasm (Quamme & Stushnoff 1983). Cell lines resistant to stress caused by polyethylene glycol have been identified for Douglas fir (Leustek & Kirby 1990) and some crop species (Tal 1983). Stability of tolerance to water stress induced by mannitol after regeneration of plants and subsequent culture of callus explants has been demonstrated for a *Prunus* hybrid (Ochatt & Power 1988), but correlations with whole plant responses to drought stress are not available.

Cross Tolerance to Stresses

Several reports exist of plants or cell lines selected for tolerance to a particular stress displaying elevated tolerance also to other stresses. Plants and cell lines selected for resistance to one herbicide are often cross-resistant to other, sometimes unrelated, herbicides (Hughes 1983). Cell lines selected for high temperature tolerance were tolerant also of increased water stress in cotton (Trolinder & Shang 1991), and high salt levels in tobacco (Harrington 1989). Tobacco cell cultures selected for cadmium tolerance displayed tolerance to heat shock and exposure to cold (Huang & Goldsbrough 1988). Selection for salt tolerance in callus of sorghum yielded some cell lines resistant to insects (Isenhour 1988). Although an underlying genetic basis has yet to be demonstrated in these particular studies, such examples of cross-tolerance suggest the existence of common factors in stress tolerance. A protective effect of proline against a range of stressful conditions has been demonstrated (Tantau & Dorffling 1991).

Selection for Other Traits

Resistance to amino acid analogues has been used to identify genotypes producing increased levels of certain amino acids in which plants are often deficient, and stability in regenerated plants and their progeny has been demonstrated in some cases (Widholm 1988). Tobacco cell lines synthesizing high levels of nicotine have been selected on the basis of resistance to nicotinic acid (Robins *et al.* 1987). Variation in the production of the antihypertensic alkaloids serpentine and ajmalicine by both plants and callus cultures was examined among 20 genotypes of*Catharanthus roseus*. Poor correlations suggest that accumulation in callus cultures will not be a useful method for selecting plants producing increased levels (Roller 1978).

The gene determining parthenocarpy in tomatoes is recessive and difficult to detect under field conditions, but can be distinguished readily on the basis of ovary size when cultured on media with gibberellic acid (Young 1990). Higher heritability under *in vitro* rather than field conditions is also the basis for *in vitro* screening for the presence of bracts and pink discolourations (both undesirable traits) in cauliflower (Kalia 1986). Ethylene-resistant cultivars of *Begonia* display greater "keepability" under house conditions than ethylene sensitive cultivars, and an *in vitro*selection system has been developed involving exposure of plantlets to ethylene and retention of those retaining green foliage (Hvoslef-Eide 1991). Although cell lines vary for growth rate in culture, this parameter is poorly correlated with plant growth in the field (Widholm 1988), preventing selection for yield on this basis.

In summary, many recent publications have reported useful correlations between *in vitro* responses and the expression of desirable field traits, most commonly for disease resistance although some positive results are available also for tolerance to herbicides, metals, salt and low temperatures. These reports concern selection among variants (at both the sporophyte and gametophyte level), originating from sexual recombination, and also those purportedly induced in culture. Least ambiguous are reports of assays involving genetic variation generated prior to culture, and these are most numerous for disease resistance. Successful *in vitro* assay procedures have been reported for many plants and diseases (both fungal and bacterial), and the potential clearly exists for developing protocols for others. The majority of reports of *in vitro* selection concern selection among populations of somaclonal variants, and these should be treated with a little more caution. There is little evidence to suggest that *in vitro* selection for growth traits, or tolerance to high temperatures or moisture stress, would be possible.

APPLICATIONS IN FOREST TREE IMPROVEMENT

Like any other method of indirect selection, the value of *in vitro* selection is dependent on:

- The extent to which *in vitro* and field traits are correlated. This also will incorporate juvenile- mature correlations where juvenile material is used for the *in vitro* test. It should be noted that many important characters operate at above the cellular level, e.g. resistance due to phenological features, and these won't be expressed in cultured cells. Other traits may be cellular, but not usually expressed in tissue cultures.

- Heritabilities of both *in vitro* and field traits.
- Ease (and thus cost) of measurement of both.

As discussed above, correlations between *in vitro* and field responses have been reported for many diseases of plants. Traditional methods of screening for disease resistance involve field and greenhouse trials. In many of these trials, the level of exposure to the pathogen is difficult to control, and heritability is low. *In vitro* methods offer advantages, and clearly there is some potential for broader application in crop breeding. For forest tree species, on the other hand, disease resistance is not a selection criterion of widespread importance. The application of *in vitro*selection for disease resistance is therefore limited, although there may be some applications beyond loblolly pine and the poplars. Cold tolerance is an important selection criterion in some forest tree species, and exposure in the field is difficult to control. Other cold tolerance assays exist, however, which are fast and efficient and which do not involve *in vitro* selection. The leaf disk conductivity method, for example, has been shown to correlate well with whole plant performance in *Eucalyptus* (Tibbits *et al.*1991).

Tolerance to salt and metals is important in some land rehabilitation programmes, and herbicide resistance may be of use in a few programmes. In general though, greenhouse trials are reasonably simple and effective for the assessment of these traits. Decontamination of material from seedling populations for *in vitro* screening, on the other hand, is likely to be laborious. Some application is likely though where selection by greenhouse testing of candidates or their progeny is difficult by virtue, for example, of difficulties in exposing plants evenly to the agents, or in obtaining progeny of field candidates. The expense involved in decontaminating large seedling progenies for *in vitro* testing could be avoided if genetic variation could be reliably induced in culture - millions of variants in a cell culture could be rapidly screened. Applications of *in vitro* selection in plant breeding would be broader if the induction of somaclonal variation were demonstrated to be a reliable method for creating genetic variation, and if selection could be coupled with a somatic embryogenesis system.

Results of very limited studies conducted offer encouragement that further investigations of microspore selection for tolerance to low temperatures and other stresses, and the development of technologies for using the selections in breeding programmes, would be useful. As pointed out by van den Bulk (1991), a large portion of the genome is transcribed and translated during pollen development. A particularly interesting example has been the selection of *Eucalyptus gunnii* pollen surviving exposure to cold. This pollen was capable of germination and could be used in control pollinations (Boudet & Marien 1988). For the selection criteria of major general importance in foresty, in particular vigour, stem form and wood quality, however, poor correlations with field responses will limit the usefulness of *in vitro* selection. *In vitro* selection is therefore likely to have very limited application in forest trees species - of possible interest in a few programmes where there is a disease resistance selection problem, but of no broad strategic value as a research objective.

SOMACLONAL VARIATION

PRINCIPLES AND EXPERIMENTAL ACHIEVEMENTS

The term "somaclonal variation" has been used to refer to the variation which has been observed in plants regenerated from cell and tissue cultures, but in general not from axillary bud or shoot tip cultures. As pointed out by Karp (1989), somaclonal variation is a phenomenon of broad taxonomic occurrence, reported for species of different ploidy levels, and for outcrossing and inbreeding, vegetatively and seed propagated, and cultivated and non-cultivated plants.

Characters affected include both qualitative and quantitative traits. Pre-existing intercellular variation may be a contributing factor in some cases, but many instances clearly involve variation induced during culture. At the molecular level, evidence has been presented for the involvement of many phenomena, including gross alterations in chromosome number and structure, point mutations, mitotic recombination, and the amplification, deamplification, deletion, transposition or methylation of DNA sequences in nuclear, mitochondrial or chloroplast genomes (Larkin 1987, Evans 1989, van den Bulk 1991, Brown 1991, Karp 1989). Repetitive sequences may have a general role in somaclonal variation. Anomalies which seem peculiar to somaclonal populations are the occasional occurrence of homozygous variants (perhaps due to mitotic crossing over), and instances of directed changes resulting in population shifts (Larkin 1987, Karp 1989). Examples of the latter include reports of whole population shifts towards resistance to diseases caused by agents such as *Helminthosporium sacchari* in sugarcane and *Verticillium albo-atrum* in alfalfa (van den Bulk 1991). In general though, the frequency of variants displaying a desired trait is very low. The chromosomal or molecular changes may result in stable alterations which are transmitted to sexual progeny, but variants frequently are not stable, particularly through meiosis.

Some of the unstable changes may be due to the activation of transposable elements, while some may have a basis in DNA methylation (Karp 1989). Many instances of selection of variant cell lines concern epigenetic variation - adaptations to selection pressure which are reversed when the pressure is removed. The incidence of somaclonal variation is influenced by genotype, by ploidy level (polyploids giving rise to greater variation), tissue source, culture time and procedure (Larkin 1987, Karp 1989). Contrary to some early suppositions though, protoplasts are probably not more predisposed to somaclonal variation than cells (Larkin 1987). Traits with respect to which somaclonal variants have been detected include:

Disease Resistance

Variation induced with respect to resistance to disease, including fungal, bacterial and viral diseases, was well reviewed by Van den Bulk (1991). That author's listing of crops/diseases for which variants displaying increased resistance have been selected is reproduced below in Table below.

Table. Somaclonal Variants for Disease Resistance

Crop	Pathogen or disease
Rice	*Helminthosporium oryzae, Xanthomonas oryzae*
Wheat	*Helminthosporium sativum, Pseudomonas syringae*
Barley	*Rhynchosporium secalis, Helminthosporium sativum*
Oats	*Helminthosporium victoriae*
Maize	*Helminthosporium maydis*
Sugarcane	*Helminthosporium sacchari, Sclerospora sacchari, Ustilago scitaminea, Puccinia melanocephala,* Fiji disease
Tobacco	*Phytophthora parasitica, Pseudomonas syringae, P. solanacearum,* tobacco mosaic virus
Eggplant	little leaf disease
Potato	*Alternaria solani, Phytophthora infestans, Streptomyces scabies,* potato virus X, potato virus Y, potato leaf roll
Tomato	virus*Fusarium oxysproum, Pseudomonas solanacearum,*
HOp	tomato mosaic virus
Alfalfa	*Verticillium albo-atrumVerticillium albo-atrum, Fusarium solani, F. oxysporum*
Celery	*Fusarium oxysporumFusarium oxysporum, Septoria apii, Cercospora apii,*
Lettuce	*Pseudomonas cichorii*
Rape	*Bremia lactucae,* lettuce mosaic virus
Banana	*Phoma lingam, Alternaria brassicicola*
Peach	*Fusarium oxysporum*
Poplar	*Xanthomonas campestris, Pseudomonas syringaeSeptoria musiva, Melampsora medusae*

A forest tree species not included in the above review is *Larix decidua,* for which somaclones resistant to *Gremmeniella abietina* have been identified, although performance of regenerated plants was not reported (Skilling *et al.* 1983). The majority of the citations in Table 3 involve selection at the whole plant level, after regeneration from culture. Many also involve *in vitro* selection. For about half of the cases reported, resistance has been inherited in a stable manner. Mode of inheritance has been established in only a few cases, but Mendelian, quantitative and maternal inheritance have all been implicated (Van den Bulk 1991). Heritability of resistance has not been established in many cases. In some cases, it has been shown that plants selected for resistance were changed in other, unwanted traits as well. Many reports exist also of unsuccessful attempts to select regenerants with increased resistance.

Herbicide Tolerance

Because of the simplicity of adding herbicides to culture media, herbicide resistance was an early target of somaclonal research. The review by Hughes (1983) demonstrates that cell lines and/or plants have been selected for resistance to most types of herbicide:

- Carbamates (*Brassica campestris,* tobacco, soybean)
- Dipyridylium compounds (soybean, tomato, tobacco)
- Glyphosate (tobacco, carrot, potato)
- Phenoxy compounds (*Trifolium repens,* tobacco, carrot)
- Picloram (tobacco)
- Atrazine (tomato)
- Amitrol (tobacco)

Plants regenerated from cultures selected at the cell stage may or may not be resistant. In greenhouse trials, somaclonal variants of poplar hybrids have demonstrated tolerance to glyphosate and sulfometuron methyl (Michler and Haissig 1988), in the latter case at ten times the concentration required to kill the original clones (Michler 1988). Sexual transmission of the herbicide resistance has been demonstrated for picloram (tobacco), paraquat (tobacco and tomato), amitrole (tobacco) and chlorsulfuron and sulfometuron methyl (tobacco) (Chaleff 1986b). Where investigated, herbicide tolerance often has been shown to be a genetically simple trait - dominant and semidominant nuclear allelles at one to three loci for resistance to paraquat in tomato and to picloram, chlorsulfuron and sulfometuron methyl in tobacco (Chaleff 1986b). Atrazine resistance, evident in plants regenerated from resistant cell lines of soybean, is not passed on to progeny as a simple nuclear dominant trait, but may be cytoplasmically inherited (Wrather & Freytag 1991). Plants regenerated from *Nicotiana debneyi* calli tolerant to amitrole displayed some tolerance, but transmission of the resistance trait to F_1 progeny did not seem to display either simple Mendelian patterns or maternal inheritance (Swartzberg *et al.* 1985).

Tolerance to Salt and Other Stresses

Once again, technical simplicity has meant that identification of salt tolerant variants has been a popular subject for research. In some cases, somaclones displaying tolerance stable through subsequent sexual generations have been identified, e.g. in *Brassica juncea* (Jain *et al.* 1991) and alfalfa (Winicov 1990), as a dominant mutation in the latter case (Winicov 1991). In several reports, salt tolerant plants have been regenerated from selected cell lines, although stability through meiosis is yet to be demonstrated, e.g. for *Prunus* (Ochatt & Power 1989) and *Citrus*(Spiegel-Roy & Ben-Hayyim 1985). Many reports though concern selection for tolerance which is not stable or the stability of which has not been demonstrated.

Aluminium tolerance was expressed in *Nicotiana plumbaginifolia* plants regenerated from variant cell lines selected in culture (Meredith *et al.* 1988). Distribution of tolerance among sexual progeny of the first generation was consistent with that expected for control by a single dominant gene. In subsequent generations though, this clear segregation pattern was not evident. Rice cell lines selected for aluminium tolerance yielded tolerant plants (Van Sint Jan & Bouharmont 1992). For potato, 30% of tolerant cell lines yielded tolerant plants, but only 5% of these clones displayed tolerance after four micropropagation cycles (Wersuhn *et al.* 1988). These interesting examples illustrate the complexity of somaclonal variation, and highlight the importance of studies, in particular genetic studies, to determine the stability and value of variants identified. In most cases, somaclonal variants displaying tolerance to stresses have not been subjected to such studies. Tolerance to higher levels of cadmium in tobacco cell lines was stable in the absence of cadmium, but not tested in whole plants (Huang & Goldsbrough 1988). These tolerant cell lines also showed higher tolerance to heat shock and cold treatments than the unselected cells. Cold tolerant wheat plants were regenerated from cell lines surviving immersion in liquid nitrogen (Kendall 1991). Wheat cell lines selected for resistance to hydroxyproline, resistance which was stable in the absence of the agent, showed increased frost tolerance (Tantau & Dorffling 1991). Leaf explants from plants regenerated from cotton cell lines resistant to high temperatures also displayed increased tolerance (Trolinder & Shang 1991). Tissue culture derived variation for improved tolerance to acid soils, drought and insects has been reported for sorghum (Miller *et al.* 1991).

In summary, somaclonal variation has been a popular subject for research, particularly during the 1980s. Forest tree species have not been neglected in this research. Significant work has been conducted on resistance to both disease and herbicides in poplars, and to disease in *Larix decidua*. A few commercially useful cultivars have been produced for crop species. With respect to disease resistance, for example, new cultivars have been released for sugarcane (resistant to Fiji disease), tomato and celery (resistant to *Fusarium* diseases) (van den Bulk 1991). Tomes' (1990) documentation of somaclonal variants released or in field trials included chlorsulfuron and imidazolinone resistance in rape, imidazolinone resistance in corn, high solids and*Fusarium* race 2 resistance in tomato, a white flowered form of lucerne, potato virus Y resistance in tobacco, and fall armyworm resistance in sorghum. In general though, somaclonal variation has not had the agricultural impact that early results promised. A significant problem is that many different phenomena are involved in variation induced in culture. Some of these are useful, and some not. In general, somaclonal variation is still poorly understood, and certainly not controllable. Furthermore, practical application is dependent on the ability to regenerate from the cell lines. This is still a problem for many species, and somaclonal variants often show reduced regenerability even for normally tractable species. Few studies have incorporated sufficient genetic examination to evaluate the stability or usefulness of the variants selected. This is not a field where success with a new crop/trait could be predicted with any confidence. Such a research program would be very empirical with the odds against success.

FUNCTION TO FOREST TREE IMPROVEMENT

Assuming stability of the characteristics and the ability to regenerate from the variant cell lines, the use of somaclonal variation is likely to be of most interest where:

- Insufficient natural variation is available to provide the level sought for the trait; or
- Existing variation is not easily used in breeding.

Many crop species in current use show greatly reduced genetic variability as a result of a long history of selection and breeding. Peach germplasm, for example, is narrow, and regeneration of plants from selected cell lines provided resistance to bacterial spot which was not available in commercial cultivars or a large number of introductions (Hammerschlag 1986). Commercial and breeding populations of most forest tree species generally have not been subjected to the same reduction of genetic variation. The poplars are a possible exception. Even for genetically broad populations, however, there still will be traits for which insufficient natural variation is available. A survey of populations of a number of pine species, for example, revealed no useful resistance to *Gremmeniella abietina* (Skilling *et al.* 1983). Similarly, tolerance to many herbicides is not known to exist in angiospermous tree species and would thus be difficult to achieve through traditional breeding (Michler & Haissig 1988). Cold tolerance in many *Eucalyptus* species is another example. It has been suggested (Larkin 1987) that plants have present in their genomes a far richer array of genetic information than that expressed, that the phenotype represents a limited subset of the genes present, and that somehow cell culture gives access to some of this genetic resource.

A proven technique being available, somaclonal variation may be of value also where natural variation is difficult to manipulate. This approach may in some cases be less disruptive to the commercial genotype (and therefore quicker) than a backcrossing program with a wild type. For forest tree species, breeding programmes for poplars are based on crossing desirable traits, in particular resistance, into commercial clones, and a successful somaclonal approach may offer some efficiencies, particularly if combined with *in vitro* selection. In general though, somaclonal variation offers very little for the genetic improvement of most of the major industrial forest tree species - the quantitative traits of most interest e.g. vigour, stem form and wood properties, are not traits which have been usefully altered *in vitro*. Poplars, often capable of regeneration from cell lines, and for which disease resistance is an important criterion, are perhaps the only candidates worthy of consideration among these species.

No immediate applicability is evident to the other tropical hardwoods or to non-industrial species, for which genetic variation naturally available is generally poorly defined. To summarize the status with respect to developing countries, this technology has very limited application in the short or imtermediate term, and any research would be high risk (with a low probability of success). This research should not be a high priority. In the longer term, the promised benefits of somaclonal variation may be more reliably delivered by genetic engineering approaches.

BIOTECHNOLOGIES AND TREE IMPROVEMENT

The author has reviewed the current status of biotechnology, and applications in tree improvement in detail in a work soon to be published (FAO, in press). The following paragraphs summarize the most important biotechnologies.

CRYOPRESERVATION AND IN VITRO STORAGE

This comprises the maintenance of cells, tissues or organs in cultures where growth is slowed (e.g. by the reduction of light, temperature or nutrients) or suspended (by immersion in liquid nitrogen). Many technical difficulties are involved, particularly in the subsequent regeneration of plants from the cultures, but recent results are generally encouraging. Regeneration from cryopreserved tissues has been induced for more than 70 plant species, including coconut, rubber, cocoa and coffee, and for several forest tree species. These results have led to hopes that the technologies may have a number of applications in tree improvement.

Gene conservation. Although increasingly used for the storage of threatened germplasm of agricultural species (Engelmann, 1991), *in vitro* storage and cryo-preservation have little to offer for this purpose with regard to forest trees. Gene pools of most of the established industrial species are reasonably well preserved in stands, both *in situ* and *ex situ,* and in seed stores. Undoubtedly, the gene pool of many tree species is threatened, particularly among the tropical hardwoods and non-industrial species. Distributions of these species are poorly known, as are their biological characteristics. Major impediments to the preservation of forest tree germplasm are: the inadequacy of resources for the survey and collection work that would be required before any germplasm could be stored; and the unreliability of many existing seed storage facilities. Even for the recalcitrant (hard-to-store) species, priority would be better directed to the establishment of *ex situ* plantings, which should facilitate urgently needed evaluations of the material. In the longer term, cryopreservation and *in vitro* storage may have some application as a backup conservation strategy, but only for populations of well-surveyed, recalcitrant species.

Maintenance of juvenility. Suspension of the growth processes also implies the maintenance of the maturation state previously attained in the tissues - without any of the uncertainty associated with alternative strategies such as long-term hedging or serial propagation. Cryopreservation therefore warrants much more attention as a means of maintaining juvenility during simultaneous clonal testing and thus capturing genetic gains offered by clonal forestry with industrial species. The technology is therefore applicable mainly in cases where good breeding programmes are in place, where clonal forestry is a realistic goal and where "rejuvenation" is difficult -particularly for the conifers.

Use of molecular markers

The use of molecular markers involves the examination, using sophisticated biochemical techniques, of variations in cellular molecules such as DNA and proteins. As an alternative to traditionally measured features such as vigour, stem quality and various morphological aspects, molecular markers offer the advantages of being unaffected by the environment or the developmental stage of the plant while also being very numerous. These characteristics have led to a number of potential applications in tree improvement.

Genetic fingerprinting. The inherent characteristics of molecular markers render them much more useful than morphological traits in establishing the identity of a particular tree or tracing its genetic relationship to other trees. For example, using molecular markers, it was possible to identify each of 39 peach cultivars individually (Ballard *et al.*, 1992). Markers have important immediate applications in supportive research for advanced breeding programmes with industrial species mainly for quality control, e.g. checking clonal identification, orchard contamination and within-orchard mating patterns by "fingerprinting". Markers also have important immediate applications in supportive research for tropical hardwoods and non-industrial species, in particular for essential taxonomic studies and investigations of mating systems.

Quantification of genetic variation. Molecular markers are potentially more useful for quantifying genetic variation than are traits such as vigour and stem form, for which environmentally induced variation is frequently a confounding factor (i.e. it is not clear if traits are produced genetically or by external factors). Markers have been used to compare within- and inter-population variation in several tree species (Muller-Starck, Baradat and

Bergmann, 1992). The quantification of genetic variation to aid in sampling strategies for the development of gene conservation and breeding populations of new industrial and non-industrial species is a potentially useful application of molecular markers. Markers may, however, provide underestimates of genetic variation with respect to traits (e.g. vigour and stem quality) that are more subject to evolutionary pressures and, therefore, will need to be used with caution.

Marker-assisted selection. This refers to indirect selection on the basis of markers shown to be associated with commercially important genes. Unaffected by the environment or developmental stage, markers offer the possibility of highly effective and early selection, long the hope of forest tree breeders (e.g. selection for wood quality at the young seedling stage). Although the possibilities are very attractive, there are limitations that will prohibit application in the short or medium term (Strauss, Lande and Namkoong, 1992): i) marker analysis is currently too expensive to permit the screening of large populations of seedlings; ii) associations between markers and economically important traits have to be established separately for different families, thus, even when cheaper markers are available, marker-assisted selection will apply mainly to advanced and sophisticated breeding programmes - those for which the creation and maintenance of the appropriate pedigree structures can be afforded and where clonal forestry is achievable. For most species, current resources would be far better directed towards moving breeding programmes to this stage of advancement, rather than to the development of marker-assisted selection.

The major current value of molecular markers lies in long-term strategic research; marker studies are making great contributions to advances in the understanding of basic genetic mechanisms and genome organization at the molecular level. An important emphasis of this work in coming years will be the study of quantitative traits of forest trees, of which a few model species will receive most attention, for example loblolly pine *(Pinus taeda).*

Genetic engineering

This comprises the insertion of novel genes into a plant or else the modification of existing genes through manipulation of the DNA molecule. Crops to which genes for insect, virus and selected herbicide resistance has been added are being or are near to being applied commercially. A tree crop into which these genes have been inserted is the poplar. Many projects are under way for forest trees, the reduction of lignin biosynthesis, for instance, but numerous technical difficulties remain to be solved. The insertion of currently available insect- or herbicide-resistant genes into a new species would constitute a major research undertaking, and successful application would be dependent on being able to regenerate from the transformed cells. The manipulation of more complex traits would be an even more formidable undertaking and much research remains to be done. An often overlooked research component is the extensive testing that would be required before a responsible recommendation for the large-scale deployment of transgenic plants could be made. Research projects of this type are necessarily intensive and must be regarded as long-term with only a modest expectation of success.

Insect resistance is of potential value, for example in poplars and some tropical hardwoods. However, the work involved in introducing several different resistance genes, sufficient to ensure that insects do not acquire tolerance during the rotation, should not be underestimated. The reduction of lignin biosynthesis is a very valuable objective for the pulp species. The introduction of herbicide-tolerant genes is of some interest but, in many programmes, the advantage of practicing unguarded herbicide applications may not be sufficient to pay for the research programme. Cold-tolerant genes are likely to be of some commercial value for many species, in particular the eucalypts. Much remains to be done, however, to establish that sufficient tolerance can be conferred using antifreeze proteins and to extend the work to tree species.

Prevention of the escape of genes into wild populations is likely to become an important concern, and sterility should be an early target of genetic engineering work with forest tree species. The major factor limiting application of genetic engineering in forest trees is the state of knowledge of molecular control of the traits that are of most interest - those relating to growth, adaptation and stem and wood quality. Genetic engineering of these traits remains a distant prospect. It is important that genetically engineered genotypes be of high quality with respect to other traits as well. The clonal test is the most logical basis for the integration of genetic engineering into traditional tree improvement programmes. For these reasons, genetic engineering is most appropriately conducted with species for which breeding programmes are advanced and clonal forestry can be realistically contemplated. Research on this subject should not assume a high priority with species for which natural variation available within the taxon remains poorly investigated.

Micropropagation

This refers to in vitro plant propagation methods. The principal approaches are axillary budding (actually a miniaturization of propagation with cuttings): the induction of adventitious buds on non-meristematic tissue (i.e. inducing a shoot where one would not normally develop): and somatic embryogenesis (where individual cultured cells or small groups of cells undergo development resembling that of the zygotic embryo). As an alternative to other vegetative propagation methods, the attraction of micropropagation lies in its ability to multiply elite clonal material very rapidly. More than 1 000 plant species have been micropropagated, including more than 100 forest tree species (Bajaj. 1991; Thorpe, Harry and Kumar, 1991). Successful experimental practices probably could be developed for most tree species. For most industrial forest plantation species, the costs of planting stock and insufficient data regarding field performance remain major obstacles to be overcome before a broader use of micropropagules as direct planting stock may be contemplated (Haines, 1992). Micropropagation has an immediate application, however, in integrated clonal propagation systems featuring the commercial planting of cuttings harvested from rapidly multiplied, micropropagated stool plants of the selected clones. This approach is of value only in very advanced breeding programmes incorporating the identification of outstanding clones currently only a few programmes are at this level. Appropriate integration into breeding programmes is essential. Where clonal testing on a relatively large scale is possible and affordable, the currentapplicability of techniques mainly to juvenile material is not necessarily an impediment to the capture of good gains through clonal forestry. This conclusion, however, is dependent on the ability to store juvenile material for the period of a clonal test. Genetic variation in response, often substantial, is not likely to be a major problem where clonal testing can be preceded by screening for responsive genotypes, although demonstrating the absence of adverse correlations with economic traits is important. Breeding programmes new industrial species and non-industrials are not sufficiently advanced to warrant much use of micropropagation in the short term.

Micropropagation may have a wider application in the multiplication of stool plants of industrial species as breeding programmes become more advanced and other limitations to clonal forestry (e.g. maturation problems) are overcome. For some non-industrial tree species, micropropagation may ultimately have a role in the multiplication of selected varieties prior to release. Development of simple micropropagation techniques for those species for which such methods are not already available is therefore a useful research objective but should not take priority over issues such as advancement of the breeding programme. Work done with some crop species indicates the possibility of encapsulating somatic embryos to form artificial seeds which can then be handled like conventional seeds. With considerable research, developments in this area may overcome the constraint of planting stock costs (discussed above) and enable the direct use of such propagules in forest plantation establishment. For industrial species, therefore, the development of these technologies is a useful long-term research objective but one which is best pursued with one or two model species, for example *Picea abies* and *Pinus taeda*.

In vitro control of the maturation state

There have been several reports of cultured mature buds displaying a reversion to a more juvenile state in response to the culture techniques and conditions. This has led to hopes that in vitro rejuvenation may be the solution to the poor rooting and vigour displayed by shoots collected from trees of selectable age for many forest plantation species. The major limitation to this approach is that there is little evidence of complete, permanent and reliable rejuvenation. In fact, some studies have clearly demonstrated the effect to be a temporary response to the culture condition. Further empirical work with this objective has a low probability of success. An understanding of the molecular basis of maturation (e.g. Hutchison and Greenwood, 1991) is much more likely to lead to practical manipulation, but this work is in its infancy and the reversal or promotion of maturation to precise levels remains a distant prospect. For clonal forestry with industrial species, the maintenance of juvenility is about as useful as rejuvenation for many purposes (Haines, 1992) and it is probably able to be achieved using technologies such as cryopreservation or coppicing. Nevertheless, a more fundamental control of the maturation state remains one of the most valuable objectives of long-term strategic research in forest tree improvement with industrial species. Rejuvenation is most applicable to efforts where good breeding programmes are in place and where other limitations to clonal forestry do not exist.

RESEARCH PRIORITIES

Supportive research

The preceding analysis suggests that the short-term possibilities for applying biotechnology in supportive research in forestry are:

- The use of molecular markers for quality control in advanced breeding programmes with established industrial species, e.g. for checking clonal identification, orchard contamination and within-orchard mating patterns by fingerprinting;
- The use of markers in essential taxonomic studies and investigations of mating systems;
- The use of markers for the quantification of genetic variation to aid in the design of sampling strategies for gene conservation and breeding population collections for breeding programmes with "new" industrial and non-industrial species.

Strategic research

Strategic research priorities relating to the application of biotechnology in tree improvement can be grouped into three broad areas:

Long-term generic research. This is most efficient if conducted collaboratively with a small number of model species, thus avoiding the diffusion of resources and efforts. High priority should be accorded to:

- Genetic engineering for sterility - this will underlie many of the eventual applications of genetic engineering;

- The use of molecular markers and DNA transformation techniques to investigate genetic processes at the molecular level, in particular those relating to complex traits such as growth, adaptation and stem and wood quality, is of particular relevance to industrial species, but will also pave the way for the application of biotechnology to non-industrial trees;
- Molecular studies of the maturation state for industrial plantation species.

A somewhat lower priority should be given to the development of somatic embryogenesis in combination with artificial seed technology as an inexpensive method of clonal propagation.

Long-term specific research. Two high priority areas are:

- Genetic engineering of useful traits, including lignin reduction in pulp species; cold tolerance, particularly in eucalypts; and insect resistance, e.g. in poplars and perhaps Meliaceae (when appropriate breeding programmes are in place). Transformation with appropriate genes (the introduction of several genes in the case of insect resistance) may be achieved within the short to medium term (the next five to ten years) but must be followed by perhaps ten years of field testing before responsible commercial deployment may be recommended;
- Marker-assisted selection, for species where breeding is advanced and where the creation and maintenance of the appropriate population structures are feasible and affordable - it will probably be ten years before this is possible on an operational scale.

Short- to medium-term research. Areas that warrant attention include:

- The examination of genetic correlations between regenerative competence and commercially important field traits (high priority);
- The development of cryopreservation methods as a means of maintaining juvenility in advanced breeding programmes with industrial species (high priority);
- The development of cryopreservation as a backup measure for gene conservation in proven species for which breeding programmes are in existence and for which seed recalcitrance has been demonstrated (moderate priority);
- The development of simple micropropagation techniques for species where none is yet available (low to moderate priority).

METABOLITES IN IN VITRO REGENERATED PLANTLETS

Since time immemorial medicinal plants have been used in virtually all cultures as a source of medicine (Kumar 2004; Patwardhan et al. 2004). The production, consumption and international trade in medicinal plants and phytomedicine (herbal medicine), have grown and are expected to grow further in the future. To satisfy growing market demands, surveys are being conducted to unearth new plant sources of herbal remedies and medicines and at the same time develop new strategies for better yield and quality. This can be achieved through different methods including micropropagation (Dubey *et al.* 2004). It may help in conserving many valuable tree species in the process and may open new vistas in the forest biotechnology.

Oroxylum indicum (Linn) Vent. (Family Bignoniaceae) commonly known as Shivnak, Shyonak, Sonpatha or midnight horror, is a small deciduous, soft wooded tree. It is distributed throughout the country up to an altitude of 1200 m and found mainly in ravine and moist places in the forests (Bennet *et al.* 1992). Several parts of this tree contain alkaloids and flavonoids (Grampurohit et al. 1994; Chen et al. 2003) of medicinal value. The plant is used in many ayurvedic preparations widely used by people for health care. The important medicinal principles obtained from it are *shyonaka patpak* and *Bruhat pancha mulayadi kwath* (Yasodha et al. 2004). *Dashmula* is one of the best known health care products of Ayurveda. The main ingredients of *Dashamula* are procured from the roots of five herbaceous- and five tree species, *shivnak* being one of them. This species also constitutes one of the ingredients in *Chyawanprasha* (Ghate 1999; Parle and Bansal 2006). Dichloromethane extracts of the stem bark and root possess antimicrobial, antifungal, anti-inflammatory and anticancerous properties (Ali et al. 1998; Lambertini et al. 2004).

MATERIALS AND METHODS

Seeds of *O. indicum* were collected from forest areas in and around Jabalpur. Seeds were germinated on sterilized moist filter paper. *In vitro* raised seedlings were given a treatment of 1-2 min each of 70% ethyl alcohol and 0.1% mercuric chloride. The explants viz. apical buds (ApB) (0.5cm-1cm) and axillary buds (AxB) (0.7-1cm) were dissected from 15-20 days old seedlings (8cm.). Explants were inoculated under aseptic conditions on to the sterile culture medium in test tubes on MS containing 3% sucrose, and plant growth regulators (PGRs) particularly cytokinins viz. BAP (1mg/l) with additive viz. $AgNO_3$ in different concentrations (0.1, 1.0, 2.0, 4.0 mg/l). The medium was solidified in 0.7% agar. The cultures were maintained in culture tubes and conical flasks and were kept in the culture room at a temperature of 25±2ºC, relative humidity of 60-70% and a light intensity of approx. 1500 lux provided by cool, white, fluorescent tubes under a photoperiod of 16/8 h (light/dark). The effect of continuous supplementation of PGRs on direct shoot regeneration was observed up to three subculture passages each of 20-22 days. All experiments were completely randomized and repeated at least twice. Each treatment consisted of 20-25 replicates.

Secondary metabolites: The tissue culture generated plants, were hardened (Gokhale and Bansal 2009). Hardened plants were grown in soil in natural conditions for 6 months. Nature grown (1-year old) and *in vitro* regenerated hardened plants (6 months) were used for this purpose.

Extraction of plant sample: In this study 1 g each of leaves, roots obtained from 1-year old nature grown plants, callus, *in vitro* raised root and leaf obtained from 6 months old plantlets were extracted in Soxhlet apparatus for 7 h. daily for 3 days with 300 ml of 70% methanol. The extract was then filtered with Whatman's filter paper No. 1 and evaporated on hot plate at 30°C. The residue was redissolved in 50 ml water and extracted three times with 75 ml 2-butanol. The 2-butanol layer was evaporated at a temperature of 30°C on hot plate. The extract was obtained in form of a yellow colored complex, which was subjected to HPLC analysis.

HPLC analysis: HPLC analysis of flavonoids was determined by the method of (Chen et al. 2003). The fraction analysis was carried out in a Hewlett-Packard 1100 HPLC. A Hypersil C_8 RP column (150x 4.6 mm I.D.) was used at a temperature of 30ºC, a flow rate of 1.0 ml/min and with a detection wavelength of 275.5nm. The solvent system was used with gradient elution: 0-15 min 0.2% formic acid from 80 to 35 % and acetonitrile from 20 to 65%; 15-20 minute 0.2% formic acid from 35% to 10 % and acetonitrile from 65 to 90%.

Results and Discussion

Shoot regeneration from embryonic axis explant (both direct regeneration as well as through callus) has been studied previously in *O. indicum* (Dalal and Rai 2004; Gokhale and Bansal 2005;Gokhale and Bansal 2008). It can be observed that supplementation of both additives with SM (selected media) (MS media with BAP 1mg/l) resulted in high efficiency of multiple shoot induction without interference of callus. From the two explants viz. apical bud (ApB) and axillary bud (AxB) the highest frequency of shoot initiation recorded were 90.54 and 93.39, shoot number (SN) 12.25 and 18.5 (Figs.1-4 and 5-8) and shoot length (SL) 2.2 and 2.5 respectively on MSM + BAP 1mg/l + $AgNO_3$ 2mg/l. Shoots were separated from multiple shoot buds in at least three successive cycles. Effect of this combination persisted up to three cycles (Table-2). This combination of additives and PGR proved best for shoot multiplication (Figs.9-10). Drastic reduction in shoot regeneration were observed with increasing or decreasing concentrations of AgNO3. Exogenous supply of $AgNO_3$ efficiently blocked the production of ethylene (Beyer 1976). Several reports clearly demonstrate that the addition of $AgNO_3$ in the culture medium significantly enhances organogenesis (Purunhauser *et al.* 1987; Kumar and Pratheesh 2004; Pati *et al.* 2004).

Various auxins have been known to possess different potential for ethylene production as well as callus formation at the basal portion of shoots. In the present study addition of $AgNO_3$ (1mg/l) with selected medium for rooting (IBA 1 mg/l with ½ strength MS medium) produced efficient healthy root systems with proper shoot growth (Table 3) (Figs.11-13). There are some reports where $AgNO_3$ has been used to inhibit callus formation during rhizogenesis in *Garcinia mangostana* (Chongjin *et al.* 1997), *Albizia procera* (Kumar *et al.* 1998) and *Cassava* (Zhang *et al.* 2001)

The shoots could be readily rooted with a high frequency on ½ MS mediumas reported earlier. MS medium (½ strength) has been found resulting in healthy, strong and efficient root systems in the present work too.

In order to acclimatize, the plants were initially kept in distilled water in conical flasks (8 days) and were then transferred to soil:sand (1:1) in polybags covered with polythene to maintain high humidity (Parveen *et al.* 2006). Such plants were transferred to earthen pots after a period of 8-10 days, irrigated regularly and then planted in the field. The HPLC analysis of the extract of regenerated plant was carried out to check the percentage Baicalein quantitatively. A sharp peak for standard Baicalein was obtained at Rt 4.78 peak for another compound Baicalein –7-o-glucoside was obtained at Rt 6.11 (Chen et al. 2003).

Flavonoid Baicalein (4.25%) was observed in the roots of *in vitro* raised plantlets (Chramatogram-3). Whereas, the highest percentage of Baicalein –7-o-glucoside (50.29 %) was obtained from the leaves of *in vitro* raised plantlets (Chramatogram-2). Analysis of nature grown plant's leaves and root sample showed 17.49 % of Baicalein –7-o-glucoside and 6.19 % Baicalein respectively. *In vitro* regenerated plantlets have been reported to produce higher yields of active compounds. The production of high yield of secondary compounds has also been reported from callus culture, from suspension culture or by using precursor. The data generated in the present study are expected to support micropropagation of *O. indicum*. However, there is a need to carry out more advanced phytochemical studies by using different protocols of micropropagation.

8

Forest Trees from Biotechnology

Most of the biotechnologies used in applied forestry today fall into the categories of tissue culture and molecular marker applications. Advances in these two areas of biotechnology have all been logical progressions and refinements of techniques that have been used in plant and tree improvement for decades, if not centuries, to produce clonal lines or varieties. Most of the popular concern about GM plants, or about the use of any of the other techniques discussed in this article, derives from concern about products grown as food for human consumption. This will rarely be a direct issue in forestry, although careful evaluation will be necessary in the case of multi-use tree species that provide non-wood forest products (NWFPs).

The argument that the genetic changes made in GMOs are "unnatural" has been raised as a concern. It is, however, difficult to say that any particular technique in itself can pose increased biological risk. It is the gene products that may pose a lower or higher risk, irrespective of the technology used to obtain them. To evaluate possible risks fully, in-depth knowledge of the particular genetic transformation is necessary, e.g. the product produced, where the transformation occurred, what "promoter" genes were used and the reliability of the expression over time (Gutierrez, MacIntosh and Green, 1999). This may be particularly true when more than one or two transgenes are incorporated. In such cases there may be requirements for longer field testing, environmental assessments and more caution in deployment (Burdon, 1999). In addition, mutation caused by gene transfer and tissue culture (somaclonal variation) can alter the function of non-target genes in ways that might take some time to detect. These, however, are not genetic issues, but a matter of having appropriate genetic testing and evaluation programmes in place prior to the release and use of GM (or even conventionally bred) trees in managed landscapes.

ISSUES ASSOCIATED WITH SPECIFIC TYPES OF TRAITS

Herbicide resistance. Herbicide resistance in poplars is probably the best-developed GM technology in forest trees. The first concern with herbicide-resistant GM plants is evidence of the development of resistance in weed populations. The risk may be substantially less in forestry than in agriculture, as herbicides are only used for a short time and with fewer applications during the early period of plantation establishment. Furthermore, total weed control is not necessary in forest tree plantations, so there is less selection pressure for resistance to weeds. The other concern is cross-breeding with wild populations. While this is possible, it is likely that wild populations that do hybridize with GM trees would have little selective advantage unless adjacent wild populations are treated with herbicide, which is rare in forestry. The main question is then the effect on the genetic structure of local wild populations if cross-breeding does occur (e.g. in *in situ* conservation areas) and the acceptability of this risk to local resource managers. If this is a concern, reduced flowering or sterility transgenes may be required, along with herbicide resistance in GM trees.

The introduction of herbicide resistance through GM technology is likely to be the most feasible and the most frequently applied genetic modification in trees. Nevertheless, it is only likely for a few well-developed species in certain situations, such as in intensive poplar fibre farms where herbicides are sometimes an approved management tool.

Reduced flowering or sterility. Reduced flowering in forest trees may be desirable in order to steer the products of photosynthesis into wood production, rather than reproductive tissues. However, since such reallocation of resources has not yet been well quantified, the main justification for reduced flowering or sterility development is when it substantially reduces the gene flow to wild adjacent populations of the same species. This may help promote greater acceptance of intensive GM tree plantation forestry adjacent to natural forests. Although research on flowering mechanisms is under way (e.g. Skinner *et al.*, 1999), the stability of sterility-gene expression over time will have to be confirmed in field trials that reflect expected rotation lengths.

Insect resistance. The development of GM crops that are insect resistant is now common, but these crops also create some of the most complex ecological situations that need to be addressed. The first issue of concern is the possible toxicity of the compounds produced in GM insect-resistant plants when they are grown specifically for human consumption or for animal feed as part of associated food chains. Second, there are ecological concerns of cross-breeding with wild relatives and the evolution of resistance in the pest populations, as with herbicide resistance. An additional and serious problem in forestry is that the long generation time of trees allows for many generations of insect populations to challenge a resistance mechanism.

The most developed GM approach for insect resistance in forestry, as in agriculture, has been the use of genes from a natural insect pathogen, *Bacillus thuringiensis*(*Bt*). Poplars are again among the tree species in which the technology is most advanced (TGERC, 1999). Research and development of other compounds is under way to reduce reliance on the relatively narrow group of natural *Bt* toxins (ffrench-Constant and Bowen, 1999). Because of the complex ecological ramifications and public concerns surrounding GM insect-resistant plants, high levels of scientifically sound laboratory and field testing will be required.

Wood property chemistry. It is now technically possible to alter genetically the chemistry of lignin in trees for easier and more environmentally friendly pulping. Genes that are important to the pathway of lignin development in wood have been modified to produce unique wood composition in very young trees (Lapierre *et al.*, 1999). However, genes of relatively large effect on wood chemistry have also been found naturally (e.g. a major recessive gene in loblolly pine [Ralph *et al.*, 1997]), so wood properties can also be modified through conventional selection and breeding. Therefore, many of the same ecological and economic considerations need to be taken into account with or without genetic transformation technology. Two important questions that remain in developing lignin-modified varieties or clones are how economically valuable plantations using such trees would be and whether altered wood may show susceptibilities to environmental stresses. If the answer to the latter is affirmative, it is again likely that reduced flowering or sterility genes would have to be incorporated into material used on a large scale when it is important that there be minimal gene flow to adjacent wild populations (e.g. in *in situ* conservation areas).

Can genetically modified trees be safely deployed in the environment?

The main ecological concerns with GM plants is the issue of gene exchange with wild populations - an issue already being considered in forestry with regard to seed orchard seed from conventional breeding programmes. In most of the situations for which GM trees will be considered, the plantation species will be an exotic one, so gene exchange would not be a factor; but in cases where it could occur, reduced flowering or sterility is likely to be a basic requirement. Much has been and will continue to be learned from agricultural research on transgene flow from GM crops to their wild relatives. Investigations into these specific questions are under way in forestry (DiFazio *et al.*, 1999). Although exotic plantation species may have limited risk of cross-breeding *per se*, the release into the environment of a GMO in most cases poses less of an ecological risk than the introduction of an exotic species. The introduction of an exotic species is a release into the environment of an entirely different genetic constitution, rather than the introduction of a single gene into a native species. The current regulations of the Forest Stewardship Council (FSC, 1999) prohibit the use of GM trees, but suggest that "the use of exotic species shall be carefully controlled and actively monitored to avoid adverse ecological impacts". This recommendation largely runs contrary to the argument made here that the biological risks and economic realities of all forest management options need to be evaluated objectively. On the other hand, a large genetic change made to the overall fitness of a native species and released without adequate consideration of local environmental risks is not appropriate either.

Clearly, potential risks must be balanced with benefits in all types of improved forest trees, considering the deployment schemes that will be used in different locations and at different times. For example, clonal varieties of fruit trees have been used in large continuous orchards for hundreds of years, and this deployment and management approach is considered by most people to be ecologically and economically acceptable, as it presents no unmanageable biotic or abiotic challenges. In plantation forestry for wood production, clonal blocks may be appropriate in some situations, such as poplar plantations, while pure or family-mix planting patterns are likely to be appropriate for most other forestry situations. In any event, deployment strategies must be designed to minimize the risks of economic losses of the stand as well as future biological losses, such as the development of resistance in pest and disease populations (Roberds and Bishir, 1997). At a minimum, therefore, a few dozen well-tested GM clones will be necessary to meet the typical production and diversity requirements necessary even in intensive forestry programmes.

To help address many of these concerns, a number of countries have developed regulations and restrictions specifying the requirements of confined field testing that are needed before commercial release (OECD, 2000). These requirements, which are necessary to reduce biological and economic risks, will undoubtedly continue to evolve. In addition to national laws and regulations, there are also broader international agreements on biosafety, e.g. the Cartagena Protocol on Biosafety (Convention on Biological Diversity, 2000).

EQUITY OF ACCESS TO GENETIC MODIFICATION BIOTECHNOLOGIES

Private investors have taken the lead for most investment in modern biotechnology, and in so doing have also had to manage the associated economic risks. In many situations such investment risks are protected by patents. An obvious concern is, therefore, that access to GM crops and trees will be controlled almost entirely by private corporations. Agreements for the use of techniques or material could be prohibitively expensive, preventing the use of the technology where it would be of value. In the past, governments were the main investors in breeding and molecular biology research. Their role may have to expand once again in this regard, in order to provide a flow of material and information that can be developed and shared by both private and public institutions (Santos and Lewontin, 1997), as opposed to being controlled largely by private companies. Nevertheless, financial constraints, rather than available technical knowledge, are probably the largest challenge ahead for applying modern biotechnology in less developed countries, although it has been shown, e.g. by the Agricultural Biotechnology Support Project (www.iia.msu.edu/absp/), that arrangements through donations are possible.

Even if financial considerations can be addressed in the research phases of biotechnology, moving the technology from the development stage to operational reality provides a new set of technical challenges with additional expenses (Polonenko, 1999). It is likely that only a few economically important species will warrant such additional investments, and it is on these economic situations and realities that the debate should focus.

Public acceptance of genetic modification

As has been alluded to previously, in terms of biotechnology research, one of the most important economic considerations separating forestry from agriculture is that trees require substantially more time for proper evaluation under both laboratory and field conditions. Rapid generation turnover in crop species has allowed genetic modification technology to develop quickly. Although this has been desirable from an economic point of view, the ecological ramifications have been far more difficult to study, as complex experiments must be carried out over several years. Public and government acceptance of GM plants is now probably as dependent on biological risk assessment and risk issue management (Leiss, 1999) as it is on technical developments or economics alone.

For GM trees, there is clearly a need to provide scientifically sound, neutral and intelligible information to the public, so that when and if GM trees become commercially available, informed decisions can be made about their release into the environment. Foresters are generally very aware of the need for public involvement in many other areas of forest management (e.g. chemical applications in forests, game management, felling patterns and plans, allowable annual harvesting levels), so decisions concerning GM trees may simply be an additional issue to be managed in the process.

CONCLUSIONS AND FUTURE CHALLENGES

The main issues that should remain relevant to the use of any modern biotechnology in forestry, now and in the future. The following are some summary points and conclusions:

- The development time, from the laboratory to the field, for all new types of tissue culture or GM trees will be substantial, usually in the order of a decade or more. While this may provide forest scientists and managers with time to evaluate many of the issues being faced in agriculture, the economic realities of relatively long generations will continue to be a major challenge to investors in biotechnology in forest trees. It appears that genetic modification will therefore become a reality only for particularly novel and valuable traits in short-rotation species in intensively managed plantations. Herbicide resistance is likely to be the first trait for which genetic modification will become economically viable, but even this may be several years away from large-scale use.

- All modern biotechnologies require large research and development investments. The allocation of funds, through either private or public agencies, needs to achieve a balance between building scientific capabilities and knowledge and supporting more applied, well proven forestry technologies (Burdon, 1994). In this regard, the investment and use of any biotechnology needs to be assessed on a case-by-case basis, considering the specific technology in the specific ecological, political and economic environment.
- The public must be confident that regulatory frameworks are in place so that commercial development of biotechnology does not take precedence over biosafety issues. However, the development of useful biotechnology in forestry should not be put at risk with extraordinary regulatory compliance costs that are driven by unwarranted concerns.
- Government participation in the development of biotechnology appears necessary to keep some flow of material and knowledge in the public domain. Without this, the costs of using privately owned genes could seriously limit further interest and development of genetic modification technology in trees. Concerns about monopolies arise with any new products, but governments can step in to deal with substantial inequities or corporate actions that are not in the interest of society as a whole.
- Even if GM trees are not used on a wide scale in the future, research and development in GM technology will provide a great deal of information about gene action and regulation, which could be of long-term value to conventional breeding programmes.
- Other forest management decisions with potentially more serious ecological consequences, including large-scale species introductions or inappropriate use of provenances or improved trees from even conventional breeding, need to be evaluated by foresters, managers and regulatory agencies in the same way as the products of modern biotechnology are.
- Geneticists have always attempted to create genetic options for the future. If we look carefully at where these current and future biotechnologies are heading in forestry, it is obvious that the research and development will largely be used in support of advancing clonal forestry. Therefore, it is in the context of appropriate clonal forestry that most of the issues need to be examined and evaluated.

From a genetic perspective, concerns that biotechnology is "unnatural" ignore the dynamic changes in the genetic code that occur within and across species genomes through modification of transposable genes or elements by virus vectors and through mutation. This article has attempted to point out that humans have already had a huge effect on the genetic structure of many plants and animals, using many different technologies which will continue to develop and change over time. However, while those in the forefront of any technology will promote its potential benefits, in the end it will be the economic and regulatory systems of governing bodies at the national and global levels that must evaluate the technology's relevance and appropriateness.

CHANGES OUTSIDE THE FOREST SECTOR

The example in the last paragraph, which would be applicable to many other developing countries, indicates the unpredictability of changes and their impact on forestry. Even with the best forecasting tools that were available three decades ago, very few of the changes described could have been visualized. Sector analysis focused on measurable values and on estimating demand and supply, largely based on projecting historical trends. The key parameters considered in assessing the direction of forestry development were limited to population, urbanization and changes in income, supply and prices of substitutes. Fundamental system-wide changes in the economy as a whole could not be captured by such forecasting techniques; thus many changes visualized then were rather off the mark. Many intentional efforts to bring about changes in forestry had little impact, while most changes were unintentional and not necessarily the outcome of planned efforts.

Since changes in the rest of the system can alter development scenarios drastically, it is important to understand how system-wide changes take place and what may be done by foresters to take advantage of the emerging opportunities. For convenience, the changes could be categorized as economic, institutional, environmental or technological.

ECONOMIC CHANGES

Among the major driving forces of change is the increasing economic integration and interdependency of countries. The rejection of centralized planning and the adoption of liberal economic policies have opened up economies, with the market mechanism becoming the most dominant determinant of change. Some of the impacts that are relevant to forestry include:

- Opening up of local markets for imported forest products;
- Increased transnational investment in forestry, especially logging, plantation development and wood processing, largely based on perceived comparative advantages;

- Improved opportunities for marketing unique products (e.g. products from medicinal and aromatic plants, which were earlier used only locally; ethnic foods) and services (e.g. ecotourism);
- probable adverse impacts on local industries that fail to adapt to the changes and thus lose the comparative advantages that they had in a protected environment.

The impacts of globalization, which results in the easy movement of capital, technology, goods and services across national boundaries, are sometimes difficult to assess. The set of criteria previously used for measuring comparative advantage for investment in forestry (e.g. nearness to markets and raw material supplies; quality and quantity of raw material) is expanding to include very different criteria (e.g. need to reduce pollution; degree of openness of economies; barriers to trade).

Institutional changes

While globalization is increasing the interdependency of countries and societies, more pluralistic institutional arrangements are emerging (FAO, 1999). Improved access to information is changing the centre of power and authority and, as part of a larger process of devolution of administrative responsibilities with wider emphasis on participatory approaches, local communities are playing a lead role in resources management decisions. The involvement of the private sector in forestry, including forestry research, is increasing (Enters, Nair and Kaosa-ard, 1998).

Furthermore, increasing awareness of forestry issues has brought civil society to the forefront of influencing forestry decision-making (FAO, 1998a). Improved access to information, the media's new consciousness of the public and the emergence of democratically functioning and transparent institutions, as well as impartial and just mechanisms to redress grievances, have all enhanced the ability of civil society to intervene in critical issues of public interest.

Increased concern for environmental protection

Undoubtedly, the most important development affecting forestry in recent years is the increasing awareness of the environmental issues relating to forest resources management, with initiatives at all levels (including the global-level Convention on Biological Diversity, Convention to Combat Desertification and Framework Convention for Climate Change, which were initiated by the United Nations Conference on Environment and Development [UNCED], and country-level efforts to revise forest policies) giving emphasis to environmental benefits. Outcomes of this awareness include ongoing efforts to develop and refine criteria and indicators for sustainable forest management, certification and labelling, codes of logging practices and the extension of protected areas.

Technological changes

Developments in science and technology, both within and outside the forestry sector - more particularly the latter - have had a significant impact on forestry (FAO, 1998b). Advances have included enhanced utilization potential for several species, reduced raw material requirements through improved efficiency in processing, substitution and lowered emission of pollutants. Recycling has increased and, as in the case of paper, could drastically reduce the demand for wood (Abramovitz and Mattoon, 1999). Research on biological aspects of forestry, including ecosystem processes, have enhanced the tools and techniques available for sustainable management of plantations and natural forests. Genetic selection, tree breeding and rapid multiplication techniques, coupled with refined site management practices, have enhanced the potential to realize higher productivity from tree plantations.

Developments in information and communication technologies are unleashing unprecedented changes. Resource monitoring capabilities, including imaging resolution and interpretation, have improved considerably. Forest cover changes, pest and disease infestations and outbreaks of fire are more easily monitored, and further improvements that will enable wider easy real-time access to resource information are a distinct possibility in the near future. The spread of e-mail and Internet access is improving the ability to communicate and thus to share information and experience. Distance is no longer a critical barrier for interaction among people. This is altering the structure and functions of organizations, with purpose-oriented networks becoming a dominant institutional arrangement (Nair and Dykstra, 1998), while the traditional vertically structured organizations are slowly fading out (EIU, 1997).

EXPLAINING THE CHANGES: LOOKING BELOW THE TIP OF THE ICEBERG

Given the potential of the economic, institutional and technological changes, why has the forestry situation failed to improve, or even worsened, in several countries during the past few years? Why is change rapid in some countries but painstakingly slow in others? In some cases efforts to bring about economic, institutional and technological changes are not possible to sustain, as the system as a whole in some way rejects them. The adoption of change seems to be largely dependent on system-wide developments and, more particularly, structural shifts in economies.

Structural changes in economies

Historically, agriculture (including animal husbandry) has been the main source of livelihood for most people in the early stages of economic development. Economic progress is largely dependent on the production of surpluses over and above consumption, and their transformation into other goods and services through either trade or investment. As populations grow, production is increased by bringing more land under cultivation or by intensifying agriculture through yield-enhancing technologies. Under the extensification scenario, forest cover tends to decline, as most agricultural expansion takes place on forest land. In many countries, agriculture remains the mainstay of the economy, and consequently population increase implies additional forest clearance for cultivation. Shifts away from land-based activities occur largely with the emergence of other sectors.

In several countries, the discovery of oil and natural gas has led to structural shifts in the economy (Wunder, 2000), with the extractive sector starting to contribute most of the gross domestic product and foreign exchange and accounting for most of the new job opportunities. In some countries, the growth of the manufacturing and services sectors has facilitated a shift from agriculture, reducing the pressure on land, thus slowing deforestation and, in some cases, even reversing the process. Dependence on land for the production of goods declines drastically as materials-based production improves or is replaced by knowledge-based industries. The attendant impact on forest cover is evident from the historical experience of most developed countries, and recently also from that of some developing countries.

Limits to structural shifts

If structural shifts in the economy can alter forest resource use patterns fundamentally, a key question is: What is the probability of such shifts taking place in most countries in the foreseeable future? Alternatively, what are the chances that such changes may not take place at all in some countries? In a highly interlinked and competitive global economy, the economic niches are increasingly being redefined and redrawn, largely on the basis of comparative advantages. This may limit significant structural changes in the economies. For example, development of the manufacturing sector in many developing countries remains constrained by low investments, small markets or the lack of competitiveness of local production. Manufacturing in developing countries is often agriculture-based and its expansion tends to have a negative effect on forests because of increased raw material requirements. In several developing countries, the indigenous manufacturing sector has relied on highly protected internal markets. Economic liberalization is, however, resulting in major changes, and removal of import restrictions is often undermining local industries because of increased competition from cheaper, and at times better-quality, imported goods.

Although a number of countries are hoping to expand their services sectors to enhance income and employment, there are inherent limitations to such an option. Most of the high value-added service sectors, including mature sectors such as banking, commerce and shipping as well as the newer sectors such as entertainment, information, technology and finance, are centred in the developed countries, whereas most developing countries focus on services at the low end of the value-added spectrum.[1] Even in the case of the "brain-power industries", which a number of countries are hoping to use for fast-track development, most of the services occupy low value-added niches - e.g. processing of credit card bills, "screwdriver technology" such as electronics assembly, and similar labour-intensive operations that take advantage of cheap labour. The potential for generating surplus from these activities is limited and there is very little scope for significant structural shifts.

Windows of opportunity provided by the development of the extractive sectors (e.g. oil and natural gas) have often been lost through private appropriation of the benefits or through the tendency to become too dependent on a single sector (Dale, 2000). When oil or natural gas (or for that matter any booming sector that generates a significant inflow of foreign exchange) becomes an important source of income, it pushes up the exchange rates, makes other sectors non-competitive and discourages diversification of the economy (a phenomenon sometimes called "Dutch disease" because it was first noted in the Netherlands in the 1960s following the discovery of large reserves of natural gas).

BEYOND STRUCTURAL CHANGES: SOME POSSIBLE FUTURE SCENARIOS

If major structural shifts are unlikely to take place in the foreseeable future, what will be the direction of changes and what are the implications for forestry? Will the future bring a situation of stable equilibrium that will enable most countries to bring about the necessary changes within the framework of the market mechanism? There are reasons to believe that changes could take place in new directions, largely because of the emergence of new fault lines in the global social and economic system.

Fault lines, disequilibrium and new scenarios

Predictable changes take place under equilibrium conditions in which it is easy to assess and understand the changes, whereas system-wide changes occur under disequilibrium conditions. As society develops and evolves, there are always new fault lines along which major changes take place. During most of the latter half of the twentieth century, the political-economic divide between the centrally planned and the market economies was the major fault line at the global level (Thurow, 1996). This divide tended to overshadow other fault lines. The collapse of the Soviet Union and the emerging dominance of the market mechanism were the major changes during the past decade.

However, the end of the cold war has brought to the surface some dormant or unnoticed fault lines. Many of the issues, such as authoritarianism, violation of human rights and corruption, that were overlooked (or sometimes tacitly approved or tolerated) to keep countries within a given political bloc have become unacceptable. Emphasis on democratic decision-making is percolating to the subnational level, resulting in efforts to devolve administrative responsibility to local levels. The emphasis on human rights is lending support to ethnic, linguistic and religious groups, often fuelling new conflicts and sometimes undermining the concept of nation states. Market-oriented development, especially increased flow of funds, technology and goods and services, is blurring national borders (Giddens, 1998). Alliances or interest groups that cut across countries are becoming important players in the economic, social and political fields. In a less polarized world with weaker nation states, new tensions are already surfacing which will have direct and indirect impacts on natural resources management, including forestry. These tensions include:

- The increasing economic and political power of transnational companies, the emergence of a large number of interest groups based on religion, ethnicity, language, etc., and consequent emerging conflicts;
- High material consumption and deteriorating environments;
- Widening inequalities and worsening distribution of income emerging from unequal access to skills, resources, technologies and markets.

Ruptures or tensions along the above fault lines are already evident. The large multinational companies, which are now consolidating their position through mergers, acquisitions and improved technologies, will face increasing pressure. In many cases, future conflicts will be between very active, highly networked organizations, representing the interests of a multitude of diverse groups, and economically powerful multinational corporations.[2] The rapid growth of information and communication technology is accelerating the process of change along the fault lines, with the increasing emergence of groups and communities that have a strong local base but interact in a transnational environment. There may also be strong local groups that are excluded from or unable to take advantage of the market mechanism. While the market mechanism has triumphed in recent years, it will have to adapt to a situation where new non-market systems are becoming an integral part of the economic and institutional framework.

PREPARING FORESTRY FOR CHANGE

Since there is uncertainty about the precise nature of future changes, it is difficult to indicate how forestry today should gear itself to adapt to those changes. Although in many situations the traditional problems will persist for some time, it is imperative that foresters and forestry organizations foresee changes beyond the immediate future and are prepared to adapt to emerging situations. Some of the most important changes that could be envisaged are outlined in the following sections.

Changing role of organizations

National-level objectives, programmes and plans will become less relevant as local communities, subnational entities and the private sector become the dominant forces in forest resources management decision-making. It is quite likely that forestry as a distinct sector or profession will cease to be relevant and will be integrated into a much broader framework encompassing all natural resources sectors. The role of national entities, especially hierarchically structured organizations such as forest departments, will diminish considerably. The ability of traditional forestry organizations to influence forest management - in terms of knowledge, resources, acceptance by stakeholders, etc. - has already declined considerably. Possible future roles for such organizations will be as impartial arbitrators to facilitate resolution of conflicts among the large number of emerging players, and as apex bodies to develop standards for the various practices that are acceptable to the diverse players. Since most of the productive functions will be performed by farmers, communities and private enterprises, the public sector is unlikely to have any major role in wood production. At most its domain will be limited to managing public goods and services, often under contractual arrangements.

New skills and approaches

Environmental issues, especially those relating to protection of watersheds and biodiversity, are becoming more prominent. Although ensuring the sustain-ability of public goods supply will still be a public domain responsibility, the approach will have to be very different from what it is today. Those who have to refrain from their usual activities to ensure the flow of environmental services for others will require appropriate compensation. Improved technologies for resource monitoring and fine-tuned management practices will have to be developed to ensure that all parties comply with their obligations. While the criteria and indicators for this will be developed at the national level, implementation and monitoring will take place at the local level. Substantial improvement may also be required in negotiation skills and conflict management at the local level, and highly transparent systems of resources management will have to be implemented.

RECENT AND FUTURE INNOVATIONS AND ADVANCES

Sawmilling and various machining processes

The most notable recent and expected advances in sawmilling and machining concern productivity gains obtained through mechanization, boosted by the introduction of computer-supported manufacturing processes. In traditional sawmilling, human judgement was used to obtain the best possible yields from the material. With modern techniques, cutting patterns are optimized on the basis of the qualitative and quantitative collection of data concerning inputs, which is made possible by the development of increasingly refined sensors of form, defects and irregularities. The data are then analysed in terms of the demands to be met, the state of stocks upstream and downstream of the work position, and priorities based on the product's rate of return. Products are produced following specific client orders in a relatively short delivery time, with stocks kept low. The development of these techniques allows considerable improvements in product quality and business competitiveness.

On-line control of sawmilling operations has recently been developed, often using visible, infrared or X-ray imaging techniques. For example, sawmilling machinery is increasingly equipped with sensors, such as lasers, that take account of the tool's behaviour or the blade's passage through the wood in order to assist and optimize manual control. With the recent arrival of new tools resulting from dynamic analysis and acoustic control techniques, the development of very economical sensors can be envisaged for the structural classification of woods, and even for the detection and pinpointing of certain defects or irregularities. Thus it is possible to foresee quality control based on classification by a variety of criteria such as size, the number of acceptable defects, colour, aesthetic properties and dryness. The extent to which these various criteria will be considered depends on the size of the company, the type of raw material and the market sector. Woods with a high level of irregularities (e.g. knots in softwoods) have already benefited from the application of this technology, which is becoming widespread in countries where the softwood industry is predominant. Processing of conventional tropical hardwoods can benefit only marginally from this technology because these woods are not subject to major defects such as knots. However, the technologies could be of great interest in tropical regions for the processing, grading and quality control of timber from plantations, such as eucalypts or tropical pines.

Further developments are expected mainly in the areas of data and information gathering on the milled product in both primary and secondary processing, and in the area of interactive quality control of the process through analysis of the main parameters, such as surface quality or precision of sawing. It will also be possible to reduce significantly the noise pollution produced by saws, while changes in these sounds can be used to determine the condition of the blades and possibly to guide cutting parameters such as feed speed. In surface machining (planing, moulding, etc.) there is a trend towards very high-speed machining, which can give a surface quality similar to that produced by sanding, while speeding up the work. Until recently, the rotational critical speed, related to the first mode of blade vibration, seemed to be an insurmountable obstacle for turning machinery, but rotation speeds between the first and second critical speeds can now be envisaged, thanks especially to advances in tensioning and computer-run saw-guides.

Complex multi-axial machining systems carrying out multiple simultaneous operations, such as the "work centres" available in the furniture industry, are gaining ground, as a result of developments in felling equipment, the widespread use of diamonds and the increased use of ceramics. These new cutting materials make it possible to adapt the tool's wear resistance to the high cutting speeds of such equipment and lengthen the use between sharpenings.

In veneering, automatic log-centring devices allow gains in time and, hence, in material. As in sawing, vibration or sound sensors will allow interactive control of veneering machines based on analysis of their performance. Furthermore, peripheral drive has made it possible to use very small-diameter logs and to reduce peeler cores to a few centimetres. On-line control systems, defect sensors, etc. also allow optimization of certain stages, such as wet clipping while the product and its various components are monitored during manufacture. The wear resistance of the steel used for blades and pressure bars will be improved by nitriding (the deposition of thin, extremely hard layers, possibly using plasma torches) and associated techniques. Research on jet cutting techniques (using lasers, water jets, etc.) will very probably remain marginal in the timber sector. The use of lasers is likely to be of interest only in the case of complex forms of machining (cutting rather than planing) or non-edge cutting in thicknesses of less than 40 mm. Conventional machining systems using chip removal methods are still by far the most cost-effective for all straight-edge cutting, and are likely to remain so.

Drying

At present, the most widely used drying technique is based on temperature and humidity control with forced ventilation. The low-temperature kiln dryer, which saw its hour of glory in the 1970s, has been almost abandoned as being too slow and entailing high risks of deterioration for fragile woods. High-temperature drying (at over 100° C), developed mainly for softwoods and some lighter hardwoods, tends to be unsuitable for solid hardwoods. The following technologies are now being developed for drying hardwood species that present difficulties and dry particularly slowly:

- Vacuum drying with superheated steam - the most effective of Vacuum technologies;
- High-frequency heating followed by a vacuum cycle - particularly recommended for thicker pieces.

Advances in modelling will allow the development of more effective and, above all, more reliable control methods. Continuous measurement of changes in the moisture level of wood, above the fibre saturation point, during the drying process is still a difficult technical problem. The absence of a reliable method of measuring high moisture content rates, other than weighing either the wood stacks or test samples, under industrial conditions considerably limits the effectiveness of the most recent generation of monitoring techniques, especially during the first steps of the drying process. Ultrasound and high frequency are among the avenues currently under research.

Wood treatment and preservation

The environmental impact of existing preservation processes has considerably hampered the development of new chemical approaches. The European directive on the use of biocides will further limit both the number of active substances used in preservation and their fields of application. Over the last decade, many pesticides have been removed from the market, and industrialized countries have banned some well-known families of products such as creosotes and pentachlorophenols. There is also much debate on aqueous-phase heavy-metal-based products of chrome or copper combined with arsenic or boron, which are used to improve durability. Since there is no real substitute for these products, a definitive ban could put an end to the outdoor use of the species in most common use today.

Solvents used as a medium for active materials have been developed considerably, and emulsion systems have been refined for non-water-soluble substances. In addition to the development and refinement of preservative products with a low environmental impact, many lines of research on the margins of preservation chemistry are being pursued and are likely to find alternative solutions. These include the following:

- The development of construction systems favouring "passive" preservation (i.e. systems that eliminate permanent contacts between water and wooden structures) and/or using naturally durable species, of which the majority are tropical;
- The use of natural insect-repellent or anti-appetant substances from tree species that have natural resistance to certain insects, particularly termites;
- The combination of durable with non-durable woods in new materials or products;
- Very high temperature treatment using hot air or a heat-conducting fluid, possibly followed by a treatment bath;
- Chemical processes such as acetylation to make wood hydrophobic, polyethylene glycol treatment and grafting of inert molecules on to the hydroxide bonds of cellulose;
- Resin impregnation procedures followed by accelerated polymerization and possibly preceded by thermo-plastification deformation cycles.

The most advanced and probably the most innovative lines of research concern the use of:

- Bacteria capable of destroying certain insects or fungi;
- Hormones that upset pest growth factors, especially insect ecdysis.

The identification of genes responsible for the natural durability of particular species could, in a more distant future, make it possible to modify the intrinsic durability of certain species through genetic modification. Such advances will depend to a large extent on the international debate on genetically modified organisms (GMOs).

Gluing and coating products

The development of chemical products has been affected by the need to take account of the effects of volatile organic compounds on human health and the environment. In the glue sector, the main innovations have concerned the development of low-formol products, in line with changing regulations. The sector now has an extremely wide range of specific products that meet both the technical demands of use conditions and the demands arising from issues of industrial implementation. Available products range from hot-melt to hot-hardening glues, with a whole gamut of intermediate products. New developments include glues for greenwood, glues for rough sawnwood with thick joints, and electromagnetic wave-accelerated polymerization technologies (high-frequency, microwave, etc.).

In finishing, problems arising from the use of organic solvents have been partially solved by the development of products with a high content of dry extracts or by the refinement of water-based systems in the paint and stain sector. Research is also being carried out on products such as radiation-polymerizable products and powdered products that do not use solvents. The lifetime of exterior facing systems has been improved by the addition of anti-ultraviolet (UV) components, improvements in the visco-elastic behaviour of films and slowing in the physicochemical degradation of existing products.

The main handicap in the exterior use of wood, however, is that the physical appearance still deteriorates too quickly. Research has shown the primary role played by the species selected in the lifetime of a finishing system; this lifetime can be doubled by the use of certain species. Research is focusing increasingly on this subject and investigating the performance of species-product combinations. Other work suggests that, in due course, it will be possible to coat wood-based materials with metallic materials such as copper, aluminium and zinc, or with carbon-based "diamond" coating.

The main developments in the future could concern improvements in the productivity of finishing procedures through the development of single-coat products to replace the classic system of three successive coats. Regarding associated technologies, haze application systems that make it possible to obtain thin and very homogeneous coatings could come to the fore. Flat composites could see the development of powder finishing techniques. The use of electromagnetic wave devices could be expanded to speed up the drying of finishes in secondary processing industries, especially the furniture sector.

Composites and reconstituted wood

The past 30 years have seen the development of many types of board in response to two main demands on the part of the furniture and construction sectors - for thin flat materials and for large areas of board. Plywood was the first response to these requirements, but its manufacture required the use of high-quality wood (well-formed, cylindrical logs, straight trunks, etc.). Fibreboard and particle board produced from defibrated or fragmented wood do not require these qualities, and allow an optimal use of forest subproducts such as timber from coppicing or from thinnings, or residues from the processing of solid wood. Variations in the type of glue used, and in the size, shape and orientation of the particles, allow the production of board with a wide range of different properties to meet the technical and economic requirements of any application.

Research is likely to develop in two main directions:

- Prior physical and chemical treatment of particles to give them desirable properties such as regularity of size and natural durability;
- Chemical treatment to help particles stick to one another, allowing the manufacture of board without the addition of glue.

Plastic-wood composites should also see interesting developments. These allow the recovery of fine residues such as sawdust. Sawdust, when mixed with plastic in a proportion of up to 50 percent, produces materials that are mechanically more resistant than basic plastics, as well as being partially biodegradable. The normal techniques used in manufacturing reconstituted elements from offcuts or small-sized pieces continue to be developed, particularly in relatively unindustrialized countries, as alternatives to the use of waste in situations where fibre or particle composites are not a valid option.

WHAT FUTURE FOR THE TECHNOLOGICAL GAP BETWEEN NORTH AND SOUTH?

For most of the twentieth century, research and the resulting technological innovations were mainly in the hands of temperate industrialized countries. Seldom have innovations first developed specifically for tropical woods been transferred to temperate or cold-climate woods. One case that comes to mind is stellite tipping, a technique developed for metal machining and adapted by the International Cooperation Centre on Agrarian Research for Development's Forêt programme (CIRAD-Forêt), the French forest research and development organization for tropical regions. The technique has now been extended to all sawing operations and is used as widely for temperate as for tropical woods.

Tropical countries are increasingly in a position to benefit from the technological developments already introduced in the North, adapted to their needs if necessary. The emergence of a strong demand for wood as a construction material and a source of energy in certain tropical regions where there are highly industrialized wood sectors, such as Southeast Asia, has transformed the trade structures inherited from the twentieth century and will contribute to the increase of specific needs for technological innovation. The role of the emerging countries of the South will be vital in preventing the widening of the technological gap between the wood sectors of the South and the North.

THE TRANSITION TO PLANTATION FORESTRY

More than 4 000 years ago humans began to change the way in which they met their food needs by moving from foraging and hunting to crude cropping and herding and, finally, to modern agricultural cropping and livestock raising. Today this transition in agriculture is largely completed, except in a few parts of the world.

It was only in the latter half of the twentieth century that humans began to make similar changes in forestry: a transition from a primitive mode of gathering forest bounty, which is created solely by nature (old-growth harvesting) to the development of the science of wood production (silviculture) (see Figure). By the middle of the twenty-first century, the transition to tree cropping will be largely completed, and the greatest part of human wood consumption will come from planted forests, most of them intensively managed.

Through tree growing, commercial wood can become a crop, as in agriculture, to be planted, tended and harvested. Tree growing allows for a choice of location and species, as well as providing the opportunity to provide intensive management. It is only with tree planting that investments in tree improvement are justified. Improvements that result in higher yields or desired traits only make sense if those gains can be captured physically and in the market. Under an intensively managed regime, trees can be grown much faster with the desired traits. Climatic conditions for this are particularly favourable in parts of the subtropical and tropical world.

The trend towards tree planting will undoubtedly continue. It is driven by two powerful forces: economics and environmental concerns. Promising economic returns on tree planting have been realized in some locations for several decades, especially in advantageous areas of subtropical and tropical South America, Asia and Africa, where biological growth rates are high. In recent decades, large areas of previously unforested land have been converted to planted forests across the globe from the Nordic countries to New Zealand and from South Africa to South America. Although much of the activity has been in the industrial world, many of the most promising opportunities are in developing regions. Many planted forests are on land that was previously in low-productivity agricultural uses. The economic returns on planted forests, especially high-yielding intensively managed forests, are sufficient to continue to induce substantial investments in plantation forestry (Sedjo, 1999a). The general trend to high-yielding planted forests is receiving additional momentum from environmental concerns which have resulted in prohibitions on harvesting from some old-growth and secondary forests and regulations that make such harvesting more expensive. As the environmental movement continues to exert pressure for the protection and setting aside of more native and natural forest areas, less of this type of forest is available for logging and the costs of obtaining wood from these sources are rising.

New forest practice acts and codes are increasing harvesting costs in natural forests in many countries. For example, Kajanus and Karjalainen (1996) calculate that new forest policies in the Nordic countries, which emphasize the preservation of biodiversity, have increased harvesting costs. Similarly, it is estimated that the recently revised forest practices code in British Columbia, Canada, which deals with roading standards, riparian zones and harvesting practices, has increased the costs of timber harvests in the province by at least 15 percent (Haley, 1996). As harvesting costs have increased, harvests have declined in the coastal mountains of British Columbia.

A dramatic example of the reduction in harvests from national forests is the drop in the United States, from roughly 60 million cubic metres annually in the late 1980s to about 15 million cubic metres today. Pressures resulting from environmental concerns are likely to be amplified through the impacts of various efforts, such as those of the Forest Stewardship Council (FSC), to audit forest management practices or certify forest products. Although auditing applies to plantation forestry as well as natural forest management, it tends to raise the cost of natural forest management relative to plantation forestry. This is especially true where the planted forests are established on lands that were previously in other uses, e.g. agriculture.

The forces discouraging logging in natural forests are unlikely to go away in the foreseeable future. The pressures to reduce harvesting in old-growth and some second-growth natural forests will add to the attractiveness of the movement to invest in planted forests.

INDUSTRIAL WOOD DEMAND

Over the next several decades, the demand for industrial wood will undoubtedly increase, but not dramatically. Since the mid-1980s industrial wood demand has stagnated, with consumption of industrial wood remaining at roughly 1 500 million to 1 600 million cubic metres annually (FAO, 1984-2000). This levelling out can probably be attributed in part to the decline in production within the countries of the former Soviet Union as markets have replaced the earlier command and control economy. Increased recycling is probably another contributing factor. Since the decades immediately following the Second World War, the increase in wood consumption has been following a downwards trend. This trend towards stabilization has occurred in the face of substantial growth in the world's population and at a time when the global economy as a whole has been experiencing rapid economic growth, particularly in highly populated Asia.

Furthermore, if the latest United Nations (UN) population projections are to be believed, there is a good chance that by the middle of the twenty-first century the population of the planet will stabilize or even begin to decline (UN, 1998). In this context, it is difficult to see the basis for any dramatic acceleration in the rate of growth of industrial wood demand. By 2050, total world industrial wood demand will be higher than it is today, but not very much - perhaps 50 to 75 percent over 50 years (Sohngen, Mendelsohn and Sedjo, 1999).

SUPPLY

The predicted transition to planted forests would make the sources of wood supply dramatically different from what they are today. The author estimates that, at present, 22 percent of the world's timber harvest comes from what are essentially old-growth forests, while 34 percent originates in planted forests, and only 10 percent in fast-growing forests, which in 2000 consist of fast-growing exotics (Sedjo, 1999b).

Table. Current and forecast global harvests, by forest management situation

Forest management situation	% of global industrial wood harvest	
	2000	2050
Old-growth	22	5
Second-growth, minimal management	14	10
Indigenous second-growth, managed	30	10
Industrial plantations, indigenous	24	25
Industrial plantations, fast-growing	10	50

By contrast, Table 1 forecasts that by 2050 up to 75 percent of industrial wood will come from planted forests, and about 50 percent from fast-growing forests. The fast-growing forests may no longer consist largely of exotics; rapid improvement is being made with certain temperate species, e.g. poplar, so that a large fraction of future fast-growing forests may be made up of improved domestic species. However, in much of the subtropical developing world, exotics are likely to dominate.

Much of the planted industrial wood would come from high-yielding intensively managed forest operations. In some parts of the world, as is already the case in certain areas of the United States and Canada today, these operations may be called "fibre farms" and may operate on rotations as short as five to six years. However, most of the increased plantation output is likely to come from the newly created planted forests of the subtropics. Globally, fast-growing planted forests will encompass an area of perhaps 200 million hectares, or only about 6 to 7 percent of the world's currently forested area. Extensively managed natural forests that meet the auditing standards are likely to continue to provide a portion of the world's industrial wood, but largely for speciality products. Such changes have profound environmental implications.

The far more rapid growth associated with intensive management implies that huge volumes of wood will be produced from relatively small areas of land. Thus a trend towards planted forests and tree breeding does not imply, as some erroneously maintain, that vast areas of natural forests are to be replaced with planted forests. Accordingly most of the world's natural forests would remain for other purposes.

Indeed, it has been seen (for example, in the National Forest System of the United States) that increased demands are placed on natural forests for other purposes as the production of industrial wood from these lands declines. Often this change is driven by policy and regulations, but such policies would not be feasible in a world where demand for industrial wood is far outstripping supply.

Technique	Increase in yield (%)
Orchard mix, open pollination, first-generation	8
Family block, best mothers	11
Mass pollination (control for both male and female)	21

ALTERNATIVES TO WOOD

One of the challenges to industrial forestry is the contention that the world's fibre needs could best be met by substituting wood with annual fibrous plants, such as hemp and bagasse. However, the following factors suggest that the prospects for other plant fibres as alternatives to wood are likely to be small over the next 50 years.

First, using these types of fibre sources involves a number of economic and ecological problems. A major economic problem is that they have a specific peak harvesting period: the crop must then be stored and preserved until it is to be processed, and both of these functions incur cost. By contrast, timber can generally be harvested throughout the year, or at least throughout a larger portion of the year than an annual crop can, so the labour and capital equipment can be used year-round. In addition, wood tends to resist deterioration better than non-woody plants do.

From an environmental perspective, it is difficult to see how an annual crop, planted and harvested every year, can be more benign to the environment than a tree crop harvested only 20 years after planting. An annual crop generates 40 disturbances of the land over a 20-year period, compared with two disturbances for a tree crop. Furthermore, the energy required for annual planting and harvesting, and the greater amounts of other inputs such as fertilizer almost surely make such an approach environmentally inferior to tree plantation (Sedjo and Botkin, 1997).

NEW TECHNOLOGY: AN EFFECT ON COSTS

While the costs of harvesting natural forests are rising because of the effects of both availability and restrictions on current practices, the costs of intensively managed plantations tend to be falling because of new techniques, improved plants and the like. Costs can be lowered in forestry, as they have been in agriculture, through the introduction of appropriate technology. New technology can reduce unit costs by increasing the yield or the output per unit cost. Similarly, innovation can make possible the same production at reduced costs, again reducing the unit cost.

Lower costs are likely to have two effects. In the short term, return or profitability per unit will rise. Over a longer period, rising profitability is likely to increase production, thereby lowering prices to consumers and eroding some of the increased profitability. The short-term effect increases profitability, while the longer-term effect leads to lower prices for the consumer. In the context of forestry, where there is a high degree of substitutability between plantation and natural forest wood for most purposes, lowering the costs of plantation forestry also has the effect of providing a financial incentive for the industry to continue its shift away from higher-cost natural forestry.

TREE IMPROVEMENT AND BIOTECHNOLOGY

As already noted, tree improvement has few applications in the absence of planted forests. As planted forests become more common, the actual and potential applications of tree improvement using traditional techniques or biotechnology increase. Table 2 shows the extent to which various breeding approaches have been able to generate increased productivity. Biotechnology is expected to serve as a useful tool in breeding. Genetic modification will probably usually be applied to superior trees obtained through traditional techniques.

Traits that are of interest to forestry and that are likely to be enhanceable through genetic engineering include herbicide tolerance, flowering control, fibre and lignin content, insect tolerance, disease tolerance, wood density, growth, stem straightness, nutrient uptake, and cold, wet and drought tolerance. Genetic modification of some of these traits, e.g. herbicide tolerance, has already been well developed in agriculture. The introduction of genetically modified organisms (GMOs) into agriculture, and now forestry, is very controversial because of a host of real or perceived risks to health, safety and the environment. It is difficult to predict the extent to which this issue will ultimately be resolved. However, it should be recognized that, globally, biotechnology need not be an "all or nothing" proposition. As with nuclear power, it may well be that some nations will utilize the technology while others will not. Thus, industrial wood from genetically altered trees could become common in parts of the world while remaining largely absent elsewhere.

The financial incentives for the utilization of biotechnology in forestry appear to be strong. Short rotation periods, which are emphasized in high-yielding plantation forestry (where rotations are typically from six to 30 years), provide an inherent financial advantage. Furthermore, the use of biotechnology, which is currently under development, to improve fibre characteristics and/or reduce lignin extraction costs would reduce processing costs, thereby improving financial returns. To the extent that costs of establishment and/or processing can be reduced or yields can be increased without increasing costs, net benefits can be achieved.

A CONFOUNDING ISSUE

Perhaps the greatest unknown for the future of forestry is the phenomenon of global warming and its possible effects (IPCC, forthcoming). Warming could not only change the global distribution of forests (although probably not greatly by 2050), but could also have an influence on the extent of forest plantations. Should forests be used as major global carbon sinks to mitigate the build-up of carbon in the atmosphere, afforestation of large areas could be undertaken as part of the mitigation process. This activity could receive further impetus should bioenergy be used on a large scale to replace fossil fuels. Finally, wood materials sequester carbon even as they are utilized, and they have a lower energy (fossil fuel) input in their production than most other materials, e.g. steel, brick or concrete. These advantages could provide additional incentives for the expansion of planted industrial forests.

DRIVING FORCES AND OUTLOOK TO 2050

The model used here in predicting the outlook for sub-Saharan forests and forestry to 2050 is a dynamic model that relies to a large extent on developments in the rest of the world. First we assume that Africa will pass through three eras of political, economic and social growth between now and 2050, characterized first by political stabilization, then by technological growth and finally by environmental concerns. A second assumption is that the influence of socio-economic, political and technological factors will become more favourable to forestry in the future. We also recognize the possible impact of global political forces that are beyond the control of Africa.

The era to 2020 will see massive and uncontrolled deforestation in Africa because of poverty, as Africa will be faced with poorly performing economies and forests will be exploited for poverty alleviation. According to the United Nations (UN), the region's population growth rate will remain as high as 2 percent per year into 2020. FAO has reported that 16 African countries still face exceptional food emergencies resulting from civil strife, population displacement and droughts. With food crisis, investment in forestry development will remain rare, and many African countries will not be able to raise internal resources for environmental protection programmes.

External resources for forestry, in the form of aid and grants, will become more scarce and may even disappear by 2020; it appears that the end of the cold war was also the end of generosity to Africa. Debt cancellation will be minimal and will have no impact on forest management. Development bank loans could be the only way of obtaining some of the limited international resources for forestry development from now until 2020.

At present, the lending rates of development banks are not sector-sensitive; infrastructure and conservation projects are currently subjected to the same lending rates in spite of the environmental benefits of conservation projects, which are weak competitors for financial resources. While many African countries (particularly those in turmoil) are indebted to development banks, the possibilities for lower rates in favour of less competitive sectors appear dim. Thus, funding for sustainable forest management will remain a major problem into the year 2020.

The acute shortage of funds for forestry will have a highly negative impact on the environment and wood supply. This will be most clearly manifested in Africa's inability to meet its domestic needs for industrial wood. While public sector funds from government annual budgets will continue to improve in the future, they will remain seriously inadequate to bridge the gaps that have arisen through long neglect of forests. Over the next 20 years, forestry as a commercial enterprise will remain relatively unattractive to the private sector because of difficult access to land, non-liberalization of timber trade, low wood prices resulting from government price distortions and high borrowing rates.

Era of technological growth (2020 to 2040)

This will be a period of real development for sub-Saharan Africa, with advances in science and technology supporting agriculture and industrialization. Stability and good governance will begin to be seen in most parts of Africa. Food security will have been ensured in many countries, with significant development observed in terms of human resources and welfare, technology and industrialization. With industrial growth there will be less dependence on agriculture for employment. Crop productivity per hectare will have improved drastically, enabling the agricultural sector to release some lands for urban and forestry uses. Private foreign capital will probably have started flowing to Africa because governments will have privatized a larger part of their energy, water, communication and transportation sectors. From 2020, a more stable population can be expected, as well as a more open economy with a high level of trade liberalization and improved infrastructure. Regional integration for improved intra-Africa trade will be strengthened in all subregions of sub-Saharan Africa. Taxation is likely to become more effective with improvements in revenue collection systems. Interest on borrowed capital will have fallen to a single digit.

In this era, however, African natural forests will be highly degraded as a result of long development neglect and exploitation of the resources for economic growth. We expect environmental degradation resulting from forest loss to reach a peak by 2040, with flooding and erosion posing the greatest development challenge in the region. Today's national parks and game reserves will perhaps remain as the only natural forests to survive. Well-kept films may be the only records of the herds of zebras and antelopes that were once abundant in Kenyan and Tanzanian parks. Much biodiversity will have been lost because of the dryness of many African ecosystems. Wood import bills for some countries will be equal to today's oil import bills. Thus, in terms of development, those countries of Sahelian Africa that lack mineral resources will begin to fall behind the countries of moist Africa, as a result of the burden of energy and wood bills.

With significant wood shortages leading to higher prices and increased trade liberalization, private entrepreneurs and farm families will begin to grow more wood on abandoned farmlands with the encouragement of tax incentives, which governments at all levels will begin to use to stimulate conservation and timber growing. Forestry research will start to receive the support of the private sector and greater appreciation from governments. Governments will also begin to promote conservation, which will become an election issue. By 2040, communities will become less relevant as agents for forest protection, and private ownership of land will become dominant. Government forestry institutions will have fewer staff, and their role will centre more on formulating policy and initiating environmental legislation. The focus of forestry training will shift from extension to providing skills for forest land management as an economic enterprise.

Era of growing environmental concerns (2040 to 2050)

While, by 2040, the structure of African forests will have changed through the loss of many indigenous species and the introduction of exotic species, the years 2040 to 2050 will see a rapidly growing concern for the environment. With great improvement in incomes, forests will begin to serve more recreation and watershed protection needs. Africans will become less dependent on forests for their energy needs as electricity and gas become more abundant. These developments will be very positive for the environment and conservation. Small-diameter logs will have greater importance for the supply of construction timber. Forest fires will be brought under control as a result of private investments in timber growing. More international agreements will have come into effective operation, and forests will no longer be national assets but of strong regional concern. Unfortunately, this may happen only when the resource has already gone. We therefore call on governments, international organizations, non-governmental organizations (NGOs) and individuals to double their efforts to avert this situation.

ASIA AND THE PACIFIC

A Russian proverb says that when we predict the future, the devil laughs. So what if the devil laughs? Let me predict the future of forestry in the Asia and the Pacific region in the year 2050, 50 years hence, and see who has the last laugh, the devil or us foresters! By 2050, forestry will have become the victim of the social, economic, scientific and political changes that will take place in the region, and in the world, over the next five decades. The walls of communism will crumble everywhere, including in China, and capitalism will be the norm of economic activity. This, together with the spread of globalization into all countries, will have great economic impacts on forestry. The development of science through research and development will have progressed by leaps and bounds, and forestry will benefit significantly, especially from biotechnology.

Privatization of forestry will become the norm in most countries. The public sector, i.e. the forestry departments, will be responsible solely for enforcement of the law and the collection of revenue. The private sector will be repsonsible for all other functions of forestry, including the management of forest plantations, natural forests and parks. This will ensure the practice of more integrated forestry in which upstream activities, traditionally in the hands of the public sector, will be linked directly to downstream activities, which traditionally have been in the hands of the private sector. This evolution will have far-reaching impacts on the development and practice of forestry as well as on the forestry profession in Asia and the Pacific. The natural forests of the Asia and the Pacific region will be diminished in area. Tropical forests will be limited to Kalimantan and Irian Jaya in Indonesia and small pockets in Malaysia and Papua New Guinea. The Philippines and Thailand will have little natural tropical forest left. Subtropical forests will be limited to isolated reserves in Southeast Asia, Australia and the Indian subcontinent, while temperate forests will still exist in small reserves in the temperate regions of China, the Indian subcontinent, Australia and New Zealand. However, the use of all these remaining natural forests will be limited to conservation, production of water and recreation, while all wood fibre needs will be met by forest plantations. Foresters will have finally recognized that they must manage the whole ecosystem, including fauna and flora, and must not be limited to managing only trees. These areas will be managed by the private sector and the public good will be one of their most economically valuable products.

FOREST MANAGEMENT OBJECTIVES

Forest recreation will be the commodity in greatest demand, and ecotourism will have been developed to a true science. Ecotours will be so popular that forests will have to be zoned for different intensities or types of eco-activities, and specific recreational management plans will have to be developed to manage recreation within the remaining natural forests. In many countries the supply of freshwater will be a major problem, and the natural forests will also function as water catchments. These forests will be managed for the production of water which will be sold to public utilities.

The greatest forest-related income will come from bioprospecting permits and agreements. The remaining natural tropical forests of Asia and the Pacific will be opened to international bioprospecting, which will become a global industry equivalent in importance to prospecting for oil. Products that will be sought include pharmaceuticals, cosmetics and other natural products for the food industry such as natural colourings, new herbs and condiments for food. Regulations will be in place to control these bioprospecting activities through a protocol under the United Nations Convention on Biological Diversity and national legislation to ensure equitable benefits and sharing of technology.

Tropical countries in particular, where the greatest biodiversity exists, will be proactive in promoting these activities. Adequate control measures will be in place to ensure that the owners of the biodiversity also receive the benefits from these activities. Bioprospecting will be such a lucrative business that maintaining the forest for such purposes will benefit forest owners more than timber harvesting. AIDS and a host of other deadly diseases will no longer exist, because cures will have been found through such bioprospecting arrangements. All the region's wood fibre needs will be produced by forest plantations, including rubber and oil-palm plantations, which will be recognized internationally as forest plantations. These forest plantations will be managed on short rotations for the production of wood fibre which will then be processed into a variety of value-added reconstituted products. New and sophisticated processing systems using wood fibre as input will provide finished products to order. The whole production process will be automatic and controlled by computers.

Carbon sequestration will be one of the most important functions of forests, as new protocols approved under the revised Framework Convention on Climate Change will allow and promote carbon credits and trading of carbon in the international futures commodity markets; biodiversity credits and transpiration credits will also be traded in the international markets. The global demand for freshwater will exceed supply, and the trading of water credits as a commodity will be on the horizon. Certification of timber and forests will be the norm; no wood fibre or timber will be sold either locally or internationally without certification for sustainable management. A system for managing the chain of custody will be in place and practised widely. In addition to the Forest Stewardship Council (FSC), other certification bodies from countries in the South will be recognized and accepted by the international market.

THE FORESTRY PROFESSION

With the management of forests under the private sector, consultancy companies will mushroom in many countries of the South, giving new life to forestry and invigorating the profession. Forestry will have become a respected profession again, even greater in stature than medicine or information technology and engineering, mainly as a result of the demand for the outdoor life and sustainable lifestyles. The forestry profession will be governed, not only by national rules and regulations with their codes of conduct, but also by an international global forestry council under the auspices of the UN. Forestry schools will proliferate as the demand for forestry training intensifies. The Asia Pacific Association of Forestry Research Institutions (APAFRI), formed in 1995, will be the premier forestry society in the region and its mandate will have been expanded beyond research, to cover the professional practice of forestry.

SCIENTIFIC ADVANCES

Biotechnology in forestry will bloom, in terms of both institutional research and the application of research results. Planted trees will be genetically modified to meet certain needs, such as maximizing carbon sequestration, fibre production or any of a wide range of other modifiable traits. Numerous transgenic trees with many different characteristics, such as resistance to diseases and maximized growth rates, will be developed and grown in plantations - although forestry will still be a long way from producing the "ideal" tree. Growth rates of 100 m3 per hectare per year will be normal in tropical species. In 2050, a new transgenic tree that grows fast, absorbs carbon efficiently and produces edible shoots and fruits will be undergoing genetic modification to produce sap that can be used as fuel to drive automobiles without any processing. Scientists will aim to start economic production of this fuel by the year 2075. Because it will be possible to produce large volumes of wood fibre, biomass energy will be the preferred energy source for power generation, and numerous small biomass power generation plants will cover the rural tropical regions of the Asia and the Pacific region.

Nevertheless, through the development of forest plantation technology, the traditional high-value indigenous timbers of the region will also be grown in selected areas through enrichment planting of degraded forests or in forest plantations. These tropical species will be grown on a 20-year rotation, but their wood will be expensive and will cater for international niche markets. The arid and semi-arid zones of the region, for example in China, Mongolia, Central Asia and Australia, will benefit from the development of new transgenic plants and trees that can survive and grow under low moisture conditions. The semi-deserts of the past will be green with plants and trees that can withstand extreme low-moisture conditions. Some of these trees will have been genetically modified to produce strong and long adventitious roots that can seek moisture deep in the soil, and aerial roots that are able to absorb moisture from the air. The aerial roots, stems and branches will also have nodules that absorb nitrogen from the air. When they have been greened, these large expanses of land, once the eyesore of the earth, will generate enough moisture in the air through transpiration to encourage the formation of clouds and even rain. The greening of the deserts, one of the oldest human ambitions, will slowly become a reality. Economical technologies for the desalinization of seawater will allow irrigation of semi-arid and arid regions close to the sea, which will become greened for habitation and agricultural production.

Progress in the greening of the deserts will help overcome the major problem of population in China and India, which will still be the countries with the largest populations in the world. Agricultural production will increase sufficiently to meet the domestic needs of these countries for basic foods. With the land available, the technologies developed and the human resources available, the Asia and the Pacific region will have become the leading producer of food and wood in the world.

EUROPE

Fifty years is less than a complete forest rotation in most parts of Europe, so forest managers today must take actions with a mental picture of conditions 50 to 100 years in the future. Projecting present trends into the distant future is also one of the best ways of understanding the present. Speculations of this nature, however, need to concentrate more on social and economic trends than on technical forestry matters, as the former have the strongest influence on forestry practice. It is assumed here that Europe will remain peaceful and prosperous, possibly even more prosperous than it is at the beginning of the twenty-first century. If Europe were again to be subject to war or a major catastrophe, natural or human-induced (such as a nuclear explosion), then forests, as well as people, would be devastated and the first priority of foresters would become the protection or reconstruction of what remained. Such events are impossible to foresee, but they cannot be excluded entirely.

Before looking forward, it is wise to look back. What have been the main structural changes influencing European forestry over the past 50 years? Since the Second World War, perhaps the most fundamental changes have been a huge rise in general prosperity and the current transition from centrally planned to social market economies.

In the forest sector, with the exception of the years immediately following the Second World War, fellings have stayed well below increment and forest area has expanded steadily. There have been few major changes in silvicultural theory and practice: the trend towards intensive monocultures favoured in the 1950s and 1960s has been reversed in response to criticism from an increasingly well informed and environmentally sensitive public. Silviculture has returned to earlier principles, which are more cautious and less economic. Forest work has become less difficult and dangerous. Improvements in chainsaw design and increased mechanization have improved working conditions but have also made wood harvesting in many parts of Europe somewhat capital-intensive.

The balance between the main roundwood assortments has changed radically: fuelwood has become insignificant in many areas as the prices of non-renewable energy have reached historically low levels (in real terms), and logs are increasingly losing ground to pulpwood. The rise of products based on reconstituted wood has provided outlets for almost all parts of the tree. Very little raw material is wasted. Recovery of waste paper is now standard all over Europe. Recovery of used wood is starting to follow the same trend. Since the 1970s, the population of Europe (especially northwestern Europe) has become increasingly aware of environmental issues and, with growing prosperity, increasingly unwilling to accept the justification of environmental damage on economic grounds.

THE EXTERNAL ENVIRONMENT

Energy prices will be significantly higher in 2050 than they are today, and there will be systematic encouragement of renewable energies, including wood. Oil will lose its preponderance as an energy source by 2030. No single energy source will replace it, but increasingly sources will be adapted to the various uses. Many political and administrative functions will be internationalized or decentralized to a regional and local level. The nation state, although still important, will no longer be the unique focus of political power. In many areas, European Union (EU) decisions will be crucial. The EU will extend to the frontiers of the Commonwealth of Independent States (CIS) and will have a major standardizing influence in all areas of life, including the distribution of wealth between richer and poorer parts of Europe. It will also be a more democratic organization than it is today. However, on the global level, the influence of the EU and other large powers (the United States and Japan) will be counterbalanced by a rise in the influence of other regional powers. The Russian Federation will pass through further periods of extreme tension, even chaos, but will emerge as a major regional power with a reasonably competitive economy. Decline in population (by 20 to 30 percent) will be a significant problem, and hundreds of millions of hectares will be left as wilderness (except for mineral extraction and forestry).

Multinational companies and international NGOs will have great influence, economically and on public opinion. Structural change will be impossible without at least the tacit support of both groups. Stable or declining population levels, combined with economic prosperity, will make labour in Europe even more expensive (in relative terms) than it is now and will encourage mechanization and automation in every field. New technologies in the communication field and elsewhere will continue to reduce costs and increase effectiveness. There will be an enormous building boom between 2000 and 2030 in the countries of Central and Eastern Europe.

DEMAND FOR GOODS AND SERVICES OF THE FOREST

Wood-based products will continue to hold a significant part of the construction and furniture markets, provided that producers continue to innovate, market their products vigorously, keep prices competitive and maintain a favourable environmental image. Composite products with tailor-made specifications, manufactured in large, capital-intensive units from homogeneous, low-quality raw materials will dominate. Traditional sawnwood with its low processing input, irregular characteristics and high raw-material quality requirements will become a luxury or niche product. Only a few companies and regions of production will have the scale of operations, capital reserves and expertise to be serious global players in this competitive environment. In Europe, such players might include the Nordic and Baltic countries (increasingly seen as a single region); northern Spain, Portugal and southwestern France; Ireland and Scotland; Austria; Poland; and the northwestern area of the Russian Federation. These areas will be marked by very intensive silviculture and a high concentration of processing plants. Elsewhere markets for wood as raw material will be weak, and profitability (if there is any at all) low.

The demand for recreational facilities will dominate in all forests near centres of population and tourism destinations. Conflicts between user groups will intensify, with forest owners (public or private) having to arbitrate. Consumers will be increasingly sophisticated and demanding as regards both product performance and environmental aspects. For example, European consumers will no longer accept wood from natural forests - but this will be of marginal importance as all natural forests will have been permanently protected or converted to semi-natural (i.e. managed) or plantation forest, and plantation-grown timber will be much cheaper.

Energy will be the major new market as, under the influence of climate disasters attributed to greenhouse gases, governments finally use the price weapon to discourage the use of non-renewable energy sources3 and encourage wood supply and demand through vigorous actions. There will be many small wood-burning installations; wood will be almost the only energy source for heating in rural areas and a significant one elsewhere. Wood-based fuels such as ethanol or methanol will be manufactured on a large scale to replace some uses of oil.

FOREST POLICY AND MANAGEMENT

Strict standards of biodiversity conservation (protection of key habitats, wildlife corridors, etc.) will be enforced everywhere, without exception. In the intensive wood production areas, which will be owned by either forest industries or large private owners linked to the industries through contract or share ownership, specialized wood production companies will carry out most management tasks. The whole system - from plantation to harvesting, transport, processing and marketing - will be highly optimized, and all links in the chain will be in constant contact with one another. Genetic improvements and intensive silvicultural regimes will bring much improved yields and shorter rotations. Trees will be selected genetically, not only for fast growth, but also for wood characteristics such as reduced lignin for pulping. In most cases, the end use of the tree will be known when it is planted. In these areas, forest ownership and management will be profitable and there will be an active market in forest land.

Outside the intensive wood production regions, the general utility of forests to society will be recognized. Forest owners' income flows will come from wood sales, mostly to local wood energy markets (which will be closely regulated, as the markets for electricity, gas or public transport are today); they will also receive public payments to compensate costs not covered by wood sales, in exchange for significant restrictions on their freedom of choice. User fees will be seen as unfair and difficult to administer, as road use fees are today. For a very few forests with high recreation value, entrance fees (similar to those already made for motorway access in some countries) will be charged.

In areas with high recreational use, the public authorities may have to take over the day-to-day management of the forest, as small private owners will not have the requisite skills. However, in more remote rural areas outside the intensive wood production regions, most forests will hardly be managed at all, reverting slowly to a more natural state and expanding naturally on to former agricultural land. A large part of forest managers' time will be spent running public participation exercises: the profession of forester will be seen as a "people" job, not a technical one.

Climate change will modify site conditions and influence yields. Some areas will become more competitive (e.g. southern Finland and Sweden), while others will experience acute problems (e.g. Mediterranean forests threatened by fire and desertification), but overall the sector will adapt successfully. Europe will have a few "Kyoto forests" (forests established and managed specifically as carbon sinks and taken into account in the auditing process of the Framework Convention on Climate Change), mostly financed by arrangements with local or national power companies, but most of the activity linked to carbon sinks will take place in areas with better growing conditions and cheaper land.

FOREST SECTOR ISSUES

Certification will no longer be a contentious issue; either certified products will have a small niche market or (more likely) all forest products will be routinely certified. At the local level, conflicts among user groups will be the most important preoccupation. At the national and EU levels, discussion will centre on the public funding that can be made available for forest management in non-intensive wood production areas. At the international level, political discussion will focus on problems of trade policy (a "level playing field" for wood producers). It will prove difficult to distinguish payments for the management of non-intensive forests (which will be considered legitimate) from subsidies to intensive wood production areas (which will be considered, in theory if not in practice, illegitimate). As today, the main preoccupation of most forest owners will be how to cover their costs and make a profit, while satisfying all the needs of society.

LATIN AMERICA AND THE CARIBBEAN

The Latin America and the Caribbean region still has vast land areas covered by forests, but the potential represented by these forests has not yet been fully recognized and developed in most countries. What will the scenario be in 2050? This question can only be answered by first looking back. As in most parts of the world, forests in Latin America and the Caribbean were in the past considered to be an obstacle to development with relatively low economic importance. For a long time, the region had a negative balance of international trade in forest products.

In the early twentieth century, even Brazil, currently the region's main producer and exporter of forest products, was a large importer of timber. For many years the United States, Canada, Finland and Sweden exported large volumes of lumber to Brazil. Factors arising from the First and Second World Wars helped reverse this situation, but in Brazil, as well as in other countries of the region, the consolidation of the forest sector did not start until much later. Land, agricultural and forest policies developed by some countries during the 1960s were perhaps the most important elements leading to changed perspectives related to forests and forestry in the region.

Forestry policies had a substantial impact, particularly in Brazil and Chile. In the 1960s these countries developed a fiscal incentive programme to support the establishment of forest plantations. The plantations, mainly based on pines and eucalyptus, soon made available uniform and low-priced raw material, recognized as an important element in attracting the capital needed to develop the forestry industry. Land and agricultural policies led to the occupation of tropical forest areas. As a result, large volumes of high-quality and low-priced logs were made available. During the same period, the tropical timber industry flourished in Asia. Tropical timber products gained new markets, and thus opened new perspectives for investments, particularly in the Amazon basin.

The regional economic crisis of the 1980s, globalization and environmental pressures were key elements in introducing recent changes to forestry developments in the region. Chile was able to modernize and open its economy faster than other countries in the region, attracting capital and developing the potential represented by the established forest plantations. Brazil faced a long period of high inflation and economic stagnation. Lacking capital and being less attractive to international investors, its forest sector developed more slowly. In any case, these two countries were the only ones in the region to develop the forest sector (at least to some extent), mostly based on plantations.

Brazil and Chile served as models to other countries. Lessons learned from the fiscal incentive programmes in these countries were used to develop fiscal and other incentive mechanisms further. Incentives are now important instruments for the expansion of forest plantations in Argentina, Uruguay and Paraguay, and other countries have shown willingness to adopt similar models. The low sustainability of agricultural projects in the Amazon, and environmental pressures were important factors for the development of new forest policies. The process started in Brazil, where several legal instruments for further regulation of forestry activities were put in place. The process is now spreading all over the region. Based on an ample discussion, Bolivia adopted a new forestry law in 1996. Peru approved a new law early in 2000, and other regulatory mechanisms are under development.

The model adopted can vary from country to country. In Brazil, for example, the model is based on privately owned production forests, while in Peru and Bolivia forests are the property of the state and are made available to the private sector as concessions. All models, however, have incorporated the principles of sustainable forest management.

REGULATION - A KEY DETERMINANT OF THE FUTURE

Regardless of the model adopted or the type of forest, all forestry-related activities in most countries of the region are heavily regulated. In many cases decisions taken by governments are influenced by international pressure. As a general rule, regulatory measures have been increasing in forestry, while in other sectors policies for deregulation predominate in response to the globalization process. Apparently, most governments of the region have failed to put in place the proper mechanisms to make environmental concerns and development policies compatible. Low managerial capability is the main problem, and this is not likely to be solved in the next few years.

In spite of the efforts made, international cooperation has not been able to help solve this and other existing limitations. In retrospect it seems that international cooperation has been an expensive and inefficient mechanism. Governments are motivated internally and externally to increase regulations. This requires new structures to enforce the legal instruments, and little attention has been paid to the efficiency of the process. The process is augmenting, and will continue to augment, costs to governments, and is transferring costs to the private sector.

Some countries have promoted the decentralization of public administration, involving state and municipal governments in forestry, environmental monitoring and other related issues. This has been considered a way of enhancing the involvement of stakeholders and of facilitating the adjustment of regulatory and development instruments to local specific conditions. The principle is correct, but the results in most cases have not been positive. Decentralization has generally resulted in overlapping of regulations, multiplication of regulatory bodies, more conflicts and additional costs. As a result of these developments, companies and countries of the region are becoming less competitive, and lower competitiveness creates more limitations for the adoption of sustainable forest management, the final goal. Other global market players are less regulated and have a competitive advantage.

AN IMPORTANT ROLE FOR PLANTATIONS

The regulatory trend will continue in the coming years and, for some countries of the region, it will probably take at least 20 years to reverse it. By that time forest products based on native sources are likely to have lost most of their market. Plantation forest areas in Latin America and the Caribbean will continue to expand. In most countries of the region, plantation forests are heavily regulated at present, but they will be less regulated in the future. In addition, from a purely economic point of view, plantations are more productive than native forests. These factors will make plantation-based products more competitive. Technology will further help plantation forests, making it possible to produce higher-quality and more uniform material more quickly. In the future, better and cheaper wood products will be produced from plantation-grown wood.

Not all countries of the region will benefit from forest plantations. Plantations are a long-term investment, and not all countries are able to ensure the legal, political and economic stability required by investors. In addition, countries with limited local markets and poor infrastructure will have less of a chance to develop forestry in the coming years. Current developments related to forest plantations clearly indicate that in the near future the Southern Cone of Latin America will be among the most important forest-product producing regions of the world. Pinus and Eucalyptus plantations will be the main source of raw material. Eucalyptus species will gain market from tropical timbers.

Capital will continue to flow into the Southern Cone countries, mostly via private investors. Large corporations will gradually replace the existing industry. Forest ownership and industrial production will be highly concentrated. The regional market will grow faster than the global market and will be important for local producers, but the region will also become an important player in the international market. Native forests in the future will be managed mainly for environmental purposes. Native forest production areas will be reduced, as native forests are set aside for environmental reasons or because of lack of competitiveness.

More and more money from international cooperation and other financing mechanisms (carbon sequestration, debt swapping, biodiversity protection, etc.) will be invested in the protection of native forests. Several governments of the region will accept these funds and will pass and enforce laws and regulations that will continuously reduce production from native forests. This will be an easier and faster means both of solving immediate problems related to lack of capital and of achieving the expectations of the countries' populations. Governments will thus reduce social pressure and gain political stability. Future generations will judge the success of this decision.

NEAR EAST

Peace and Environmentally Friendly Governance

It is hoped that, well before 2050, regional conflicts, intergovernmental disputes and internal strife will have been amicably resolved. Coups d'état will give way to democratic processes. The only way of assuming power will be through the polls. Commitment to the environment and to sustainable development will have high priority in party manifestos and election pledges. Promises will be kept and commitments effected. With better governance, public opinion will be more influential, and the roles of civil society, NGOs, the private sector and local communities will be enhanced. All of these are likely to have a positive influence on directing more attention, efforts and resources towards the environment, particularly tree planting. They are also likely to promote greater efficiency in the use of resources for this objective. Budgetary allocations for armament, war, security and the like will be shifted to more humane and constructive uses. The bulk of national budgets and bilateral and multinational funding will be directed to reconstruction, sustainable development, environmental rehabilitation and human welfare.

THE NEW FORESTER

With the settlement of internal strife and the attainment of social stability, urbanization and automation, employment opportunities will be needed. What better source of opportunity than forestry and tree planting? Discharged soldiers and draftees in national services will be retrained and deployed, together with their earthmoving equipment, trucks, water tankers, etc., to work in reconstruction and environmental rehabilitation. They, together with forest guards and foresters in general, will be trained to forego the police mentality and to accept social fences in place of barbed wire ones. This transformation will entail radical changes in the curricula of forestry schools in order to give more emphasis to non-traditional services and products rendered by forests, especially the protection of watersheds and watercourses, desertification control and NWFPs. Greater attention will also need to be given to social forestry, to the removal of "Berlin Walls" between forestry, agriculture and horticulture and to implanting contemporary concepts such as national forestry programmes, biodiversity and sustainable forest management. The ultimate objective is to produce the awaited "new forester".

In addition to discharged soldiers, other social sectors besides those currently involved in forestry and tree planting will become involved, particularly women. Indicators of this are already evident in some countries such as the Sudan, where increasing numbers of female students are joining faculties and departments of forestry (as well as other areas of higher education). Indeed, a recent batch of forestry graduates in the Sudan was made up of 11 women and only one man. Many women, as individuals or as groups, are already forest owners or are involved in private, homestead and community forestry. If the trend continues, by 2050 it may be not the women, but the men in forestry who are seeking equal opportunity.

Sustainable forest management and conservation

As attention shifts and environmental awareness grows, factors detrimental to the environment, forestry and trees will be reversed. Decades of the practice of sustainable development will finally have brought the concept home. Sustainable agricultural production will be sufficient to meet national and regional demand, or nearly so. Urban sprawl will not encroach on arable land but will inhabit deserts and mountainsides. Environmental concern will not only stop the curtailment of forest and woodlands, but will make more land available for them. With better living standards and the availability of alternative energy sources, particularly electricity, kerosene and butane gas, the need to use fuelwood or to collect it as an income-generating activity will be diminished. So will communal grazing and transhumant livestock rearing have decreased. Instead, the forest estate and arboured areas will be expanding.

With increased awareness of forests and woodlands as common property in which all people have an interest, deliberate destruction and arson on these lands will decline. Conflicts over land use in forested areas and disputes with foresters will no longer be protested by the belligerent setting of fires - currently a serious problem in the region - as the need for forest protection gains greater recognition. Sand dunes may continue to encroach on human settlements and property, but they will be undergoing stabilization and fixation. Wildlife sanctuaries, recreational parks, green spaces, botanic gardens and arboreta will be in the process of establishment everywhere. Farms, homesteads, civic centres, roads, canals and railways will be lined by multiform, colourful and seasonally flowering trees.

The resolution of internal and interboundary conflicts will facilitate the allocation of resources and ease access for the sustainable management of common watersheds, the establishment of common shelterbelts and the harvesting of runoff from rainstorms. One of the important activities in watershed management will be tree planting and conservation. This will eventually render more and better-quality water for all purposes, including irrigated forests, green areas, shelterbelts and scattered trees. Similarly, the resources once spent on war and on reversing the consequences of war will be freed up for allocation to desalinization of seawater and recycling of excess water from irrigated fields, sewage effluents and industrial wastewater, and the use of this water for irrigating trees. The benefit from the extra water made available will be maximized through the development and adoption of improved irrigation systems. Of course, even with the resulting improvement in moisture regimes, the increased numbers of trees and shrubs in the landscape and the reduced setbacks from overgrazing and factors of vegetation removal, it is not possible to expect that the environment will revert to what it was in prehistoric times. Still, many species of flora and fauna could probably be restored, either from buried seed banks or through deliberate reintroduction, to the levels of the not-so-distant past.

NORTH AMERICA

Increased Appreciation of the Environmental Value of Forests

Perhaps the most pronounced change in North American forest management over the next several decades will be a continuation of the dramatic shift in public perception concerning the value and appropriate uses of forests. In particular, publicly owned natural forests will become increasingly valued for the environmental services they provide - notably watershed protection, biodiversity conservation and carbon sequestration - instead of just their wood and other forest product values.

Clean, reliable water is and will surely remain one of the most important products of forests throughout North America, essential not only for irrigated agriculture in the western United States and northern Mexico, but also for industrial and residential use throughout the continent. In fact, the National Forests in the United States were created in large part to reverse the deterioration of watersheds during the nineteenth century and restore them to health, a process that occupied much of the first half of the twentieth century. Today, these forests encompass some 3 400 watersheds which provide drinking-water for more than 60 million people. As the economy of Mexico continues to diversify and expand, thousands of expanding municipalities will depend on water from forested lands for both domestic and industrial use.

Forest-based recreation, already a high priority in much of North America, is likely to become even more important as per capita productivity rises and leisure time increases in all countries, and as the urban middle class in Mexico continues to grow. As species continue to be lost worldwide, forests will become increasingly valued as reservoirs of biodiversity. In North America, attention will likely continue to centre on the species-rich forests of southern Mexico and Mesoamerica. But biodiversity will also be an important factor in temperate forest management, as witnessed by current efforts to modify forest management practices along the west coast of the United States and Canada in order to protect wild salmon populations.

Finally, as the effects of global climate change become more apparent, the role of forests as both carbon sinks and moderators of climatic disturbance (such as flooding) will take on a new meaning. Implications for management objectives include not only the addition of carbon sequestration to the goals of multipurpose forest management on public lands, but also specific reforestation and forest protection projects on private lands, in response to incentives provided by carbon markets.

In order to increase the role of environmental services in forest management, at least three fundamental changes in North American forest management practices will be necessary. First, substantial cost and effort will be required to restore North American forests to ecological health. This is particularly true in the western United States, where a combination of extensive harvesting and fire prevention, however well intentioned, has unfortunately led to undesirable changes in species composition, stand structure and fuel loads, leaving many forests vulnerable to uncharacteristically intense fires and disease spread. Similar, although perhaps less severe, challenges face Canada and Mexico.

Second, innovative mechanisms that reflect adequately the full value of environmental services will need to be adopted in public policy-making and market structures. For example, healthy, functioning watersheds save local communities throughout North America billions of dollars in water filtration costs. Yet the environmental services of forests have traditionally been treated as a "free good," with inadequate recognition of their true value or the costs associated with maintaining them. This is beginning to change; New York City, for example, recently decided to invest US$1 500 million in watershed management and reforestation as an alternative to paying up to US$8 000 million for new water treatment plants. To correct the problem of undervalued environmental services, it is likely that such market-based practices as conservation easements, carbon trading and "true cost" pricing of water, recreation and hydroelectric power will become more common.

Third, increased attention will need to be given to the social dimensions of forest management. Recreational use in National Forests in the United States has grown from fewer than 20 million person-days in 1950 to approaching 1 000 million person-days today, yet insufficient effort has been given to understanding the nature and management implications of this dramatic shift in forest use. As managers adapt to changes in public attitudes towards and uses of forests, research will be needed to clarify social priorities and improve understanding of human interactions with forests. As appreciation of the importance of social and institutional arrangements grows among forest managers, it is likely that new approaches will be adopted to ensure transparency and public involvement in decision-making. Criteria and indicators of sustainability at both the national (e.g. the Montreal Process) and the management unit (e.g. certification) levels, the devolution of decision-making to local institutions, and innovative approaches to public-private partnerships are just a few examples of possible means of improving transparency and public involvement in forest management. Finally, to manage forests on a landscape scale, it is likely that new mechanisms will be developed for voluntary coordination of land management across ownership boundaries, including international borders. The presence of Mexican and Canadian firefighters in the United States during the severe fires of 2000, as well as similar assistance to both countries from the United States in recent years, is testimony to the potential of such cooperation.

WOOD AND FIBRE PRODUCTION

North American forests will not, however, cease to be important sources of timber, fibre and other commercial products. On the contrary, it is probable that North American wood and fibre production will increase in the next few decades in response to overall domestic and international demand, which is likely to grow, despite product substitution effects. However, this production will probably by concentrated increasingly in privately owned plantation forests that are specifically dedicated to fibre production, rather than in publicly owned, natural forests. One of the reasons for this trend (which is already much in evidence) to continue is the reduction of harvesting on publicly held lands because of the depletion of commercially available stocks and public concerns regarding the compatibility of logging and environmental services. The introduction of new fast-growing hybrid trees adapted to a wider range of growing environments will both help to increase the comparative economic advantage of plantation-grown wood and extend its range to new areas.

This does not mean that commercial harvesting will necessarily disappear from natural forests during the coming decades. Carefully regulated harvesting is likely to be an important tool for managing forests for multiple benefits (including restoring them to health) for decades to come. In fact, a priority will be to find new means of utilizing small-diameter wood as part of the process of restoring ecological health to Western forests with high fuel loads. In addition, small, family-owned forest plots throughout the region, particularly in the northeast and the tropics, are likely to continue to be managed for high-value wood products, among other objectives.

It seems probable that genetically modified organisms (GMOs) will have a revolutionary role in fibre production in the coming decade and beyond. Given the legitimate concerns over the wisdom of creating and deploying

GMOs, especially in wildland habitats, it seems likely that their contribution to fibre production will be focused on tree plantations and agricultural crops, combined with new processing technologies for composite materials. Perhaps even more influential will be the impact of GMOs on forests via the agricultural sector, through the concentration of crop production and subsequent reforestation of marginal lands, the extension of modified crops on to lands previously unsuitable for agriculture, or both. Finally, it is impossible to overlook the potential of GMOs either to harm forests through the introduction of novel invasive species or to be of benefit to them through the reintroduction of such species as the American chestnut and the American elm.

9

Clones in Nature and Forestry

Clones are common in Nature, and they have been used as a tool of domestication for very long time and are used in large-scale forest operations. It seems to be important to remember that to put clonal forestry into perspective, The earliest land plants lacked vascular tissue and had inefficient systems of sexual reproduction and dispersal and clonal reproduction was the typical way of propagation in nature. Clonal growth was advantageous enabling physical dominance of large areas through horizontal growth.

Clonal propagation is still important in nature for many species. Many of the commonest clones are herbaceous but the habit is also found in woody plants throughout the world. In temperate broad-leaved woodlands, root suckering is a common form of clonal growth and there are good examples in the genera Populus and Prunus. Coppice is a natural growth form that has been exploited for centuries in European forests. Betula, Carpinus, Corylus, Quercus, Salix and Tilia are all genera with the ability to self-coppice.

The use of vegetative propagation of trees as a tool in their domestication has a long history. An early use was to propagate good clones of fruit trees and this has been done for thousands of years, thus humanity has long experiences of cloning trees.

For specific cases clonal forestry is often regarded as a "new technique", and "biotech" leads associations to the public to something more radical than most cases of clonal forestry. To put things into perspective it has to be remembered that many of the techniques used in modern "seedling forestry" are also "new", and the problems and uncertainties with that is often larger than the uncertainties connected to clones. Some cases where clonal forestry has been much used are described, viz. Sugi, Eucalypts and poplars.

SUGI

The first known use of vegetative propagation for forestry purposes was in Japan with Sugi (Cryptomeria japonica D. Don, a conifer) in the 15th century. Since then reforestation with monoclones or clonal mixtures has been widely used in Japan (see Ohba 1993). Cryptomeria japonica is the most widely cultivated tree species in Japan comprising nearly half of the ten million ha of Japanese plantations. Almost half of the plants used are clones (praxis differ in different part of Japan and with different owners).

The extreme uniformity of the plantations has probably contributed to greater than average damage from typhoons and heavy snow. Attacks from Sugi bark borer and bark midge have increased in severity as the area of Sugi plantation has grown, but there is no evidence of altered disease or insect virulence related to clonal forestry, despite careful monitoring.

Clonal forestry is most common on the south-western island of Kyushu where about 100 cultivars are in use. Many foresters use only 1-3 cultivars of local origin in their plantations so that a single clone may cover as much as 2 hectares.

Allergy to Sugi pollen is a health problem in Japan as a whole, but is only a minor problem on Kyushu with its large use of clones. Cloning by cuttings for several generations has reduced the incidence of male flowers, so these plantations produce little pollen. Analogously one may speculate in that a reversion of North European birch forestry to clonal forestry in a very remote future may reduce allergic reactions with birch pollen.

Eucalypts

Clonal forestry is more common with broadleaf species than with conifers. Probably the most successful example is the eucalypt plantations in countries in tropical and subtropical climates like South America, South Africa and Portugal. The total area of eucalypt plantations today is about 30 million hectares and about half of this area is planted with cloned material.

The eucalypt plantations in Brazil cover an area of approximately 3 million hectares. A succession of eucalypt species has been grown in Brazil due primarily to disease problems. Clonal forestry started in the late 1960:s and is today the dominant form of eucalypt forestry in the country. Progeny from an Eucalyptus grandis female (male sterile) growing in an arboretum were found to be natural hybrids which were resistant to pests and diseases. The offspring of this tree are believed to be Eucalyptus grandis ? Europhylla hybrids. This knowledge was rapidly adopted by the forest industry companies and they selected clones within the open-pollinated offspring from this single tree, which means that a small number of clones originating from the same family are widely grown over large areas in Brazil. This has and is still changing with most companies having good breeding programs backing their clonal programs. The clones are deployed in huge monoclonal blocks. This situation seems favourable for a pathogen, but nothing has happened yet - despite the possible risks of a disease in a monoculture. Today the companies have become more risk-conscious and are developing breeding and clonal testing programs that include strategies for clonal diversity in space and time. A typical program is now deploying about 20 clones each year and these clones will be replaced with time as new clones are selected. The production landscape forms a genetic mosaic. New genetic material has been imported to widen the genetic base for breeding.

Poplars and willows in Europe

Vegetative propagation of poplars (Populus sp.) has a long history in Europe. Organised clonal forestry started in the beginning of the 20th century. Monoclonal plantations of poplars have become a common land use especially on river plains in southern European countries like Italy, Spain and France. Some of the best poplar clones in use were selected as early as during the first half of the 20th century. Clonal forestry with poplars is common in countries with subtropical and temperate climates like Italy, Spain, France, Belgium, USA and Canada.

Single clones of poplars have been propagated extensively. There have been disease problems and the overuse of single clones and use of large monoclonal plantations has been questioned (cf e.g. Stelzer and Goldfarb 1997 p444). Poplar has covered over 100000 ha in one country and a single clone may comprise one third of this area. The focus has been to develop disease resistance within individual clones, but these clones have proved to be highly susceptible to new varieties of disease. The disease problem seems to be growing and has caused Germany to more or less abandon poplars and other countries may follow.

Nevertheless these incidents have not provided poplar growers with sufficient incitement to focus on more diverse alternatives. It is unlikely that monoclonal cultures and the wide use of some clones are the only explanation to increased disease, but they act as contributing factors. Legal and commercial reasons often favour the use of single clones (e.g. easier to use breeders right –UPOV - for single clones than varieties, commercial demands to know that varieties are up to specifications). Even willows and short rotation forestry in Sweden are threatened by disease. Resistance is broken down within a single rotation (it is decades in between replanting even when harvesting is done at intervals of a few years). Two evaluations of the Swedish willow breeding program have rather strongly recommended an increased number of selections to be made (more than a single selection released per year). Nevertheless, the growers′ organisations still do not encourage clone mixtures.

HYBRID ASPEN IN FINLAND

A Finnish forest industry company has recently started a clonal forestry program with hybrid aspen (Populus tremula crossed with Populus tremuloides) for plantations in Finland and Estonia. So far a few hundred hectares have been planted with micro propagated selected clones. A recent Finnish doctor thesis has appeared on the subject of hybrid aspen (Yu 2001). This raises the question if something similar could be done with birch. It may be noted, however, that Yu (2001) suggested that micro propagation costs are high and that the technique needs development.

OVEROPTIMISTIC STATISTICS

It is a clear tendency that statistics reported about the use of vegetative propagation, at least there it is regarded as modern hitech biotech, gives the impression of more rapid progress than actually occurs. E.g. Talbert et al. (1993, p. 148-149) reported, based on a survey, that four Swedish operators produces 7.3 millions rooted Norway spruce cuttings annually. Lindgren et al. (1990, p. 8) estimated the annual production to 5 millions and believed in a raise to 10 millions. However, looking backwards (Sonesson et al. 2001, p. 15), less than 20 million were planted in total till 2000 and now only some hundred thousands are planted annually.

Crop uniformity

As genetic diversity and possible uniformity is an evident and possible important aspect of clonal forestry. For evaluation of possible ecological impact diversity seems central. I will thus start to discuss disadvantages with a low genetic diversity in a stand. In other respects this paper focuses on the advantages of clonal forestry.

DISADVANTAGES OF A UNIFORM CROP

There are reasons to believe that the biological production can be higher in a genetically diverse crop than in a uniform:

- A single genotype demands the same things at the same time, and thus utilises the site worse than a mixture of genotypes.
- In a mix, another genotype may take over the ecological space left by a failed genotype.
- A disease spreads faster in a uniform crop.

These statements are supported by the average of a very large number of agricultural mixing experiments and a few experiments and experiences with forestry. On average, mixes performed some percent better and were less susceptible and more stable than the average of its components, but usually the production of the mix was not better than its best component grown pure. Even if mixes seem more productive on an average, there are many examples and experiments, when no advantage was found and some experiments where diversity actually seems negative for production. I conclude that from a production point of view an advantage with some degree of diversity is expected, but that does not imply that production is higher the higher the diversity is. But the superiority in biological production is not expected to be large and it is not certain for individual cases, but just a rough generalisation. This general statement ought to apply to birch also, thus biological production is assumed to be lower and more uncertain in a uniform crop, but the effect is probably small and practically unimportant (a few percent of biological production).

Advantages of a uniform crop

There must be considerable advantages in uniform monoclonal blocks as they are used frequently. Although the advantages with diversity are repeated and reiterated and very politically correct, it is seldom applied in intensive agriculture or husbandry. The main reason is that maximizing biological production and maximizing economical production is not the same thing. It makes it easier to manage the crop if it behaves uniformly. For a forester it may not matter very much if individual birches drop their leaves on different days, but a farmer wants the crop to become ripe the same day, as the whole crop is harvested the same day. Diversity is a usually a problem for the customer, who wants a uniform raw material for further processing. For administrative processes, like describing the crop, it is an advantage if it is uniform and does not change from specification over time and space. These problems are evident for many types of agricultural crops, and therefore these crops are usually uniform even when biological production would be higher with a diverse crop.

The dream of many foresters is to find the best genotype and use only that. This has been successfully done for other domesticated species. The uniformity itself in a clonal plantation has a considerable advantage from a management point of view. It is easier in the nursery to raise plants of a single genotype. It is easier to manage and harvest a uniform forest. It has advantages to handle a single type of material. A customer or processor of a product knows more about a lot if it is a specified clone than if just the species is known. A batch of uniform logs is worth more and easier to handle than the same amount of very variable logs. Even if interest is only in quantitative production and that is lower in a uniform crop, still the higher genetic performance of the best clones compared to the best seedlings available usually much more than compensates for the production loss by a uniform crop. Thus uniform clonal crops produce more than diverse seedling crops.

My opinion is that genetic diversity often is overemphasised in comparison with the higher genetic gain achievable at the cost of a reduction in genetic diversity for future forest production plantations in many countries (like Sweden) both for propagation with seedlings and cuttings. To get genetic diversity many clones are used in seed orchards and clonal mixtures and there constraints of relatedness of clones may be seen as desirable. This was not a disadvantage at an early stage of forest tree improvement, as it does not influence gain much if many phenotypically selected plus trees are placed in seed orchards, and in the early stages of cutting propagation there were physical constraints on the number of ramets which can be produced by each clone, so it is necessary to have many clones. But now this constraint often means selection is not intensive and the best genotypes are not exploited to the extent, which would be possible. However, clonal forestry is unlikely to deal with a single clone only. That does not seem a likely scenario for birch.

Many clones will be tested and test results as well as experiences will accumulate of the best clones, so the opinion of what is the best clone will change over time. It is also dependent on the locality, the desires of the user and the availability. Even if uniform plantations are planted, there will usually be a mosaic of clones on the landscape level.

From the risk management point of view a clonal deployment philosophy offers advantages. If a clone is sensitive to a pest or pathogen the problem can be identified and managed. If silvicultural means to deal with a problem are insufficient, the stand can be salvage harvested and replanted. The clone which turns out inferior can quickly be taken out of production. It is an advantage for an operator which uses monoclone blocks to use several clones in the same time that it will become more evident what the strong and weak point of the clones are. A disadvantage with mosaics is that the uniform lots of wood from a stand are often too small to be practical as a specific variety at the end user, and actually the end user may get trouble with larger between lot differences if clonal forestry is used. Alternatively the clones can be grown in intimate mixtures, in that they most of the possible advantages of the diversity of seedlings may be gained by a mix of rather few clones.

When we are talking quality birch - and not quantity birch - the reasons for monoclonal plantings become stronger. When high quality of the product is concerned, it often becomes more important to have uniform quality also, while in a low quality product - as firewood - the quantity is usually relatively more important and a uniform and predictable product is less important. Attention on quality instead of quantity also generally coincides with a higher level of intensity, and at a higher level of intensity, clonal forestry often becomes more interesting. Thus considering the aim the project arranging this symposium has in mind; uniform clonal plantings becomes relatively more important than for forestry in general.

Improved clones can be preferable to seedlings!

There are a number of arguments for forestry with clones. It takes less time to multiply a genotype as a clone than to build a seed production unit with that clone as one of the parents and wait till that works. The time lag is actually much longer than till the first seed production, because a seed production unit is usually designed to last for many years. The risk for getting the seed storage empty means that forestry often want to have five to ten years foreseen seed need in storage. That is expensive, many seeds may never be used and the seeds actually used are less improved than possible.

A seed production unit is designed to meet a certain need of plants with certain characteristics many years ahead. If too many seeds are produced it means that the investment was unnecessary large. If too few seeds are produced it may mean that foresters must use genetically inferior unimproved seeds. If the propagation is done fast by vegetative propagation by e.g. micro propagated clones or somatic embryogenesis, it takes only a year to multiply a desirable genotype in the desired number of copies. Thus it is fairly easy to adapt to the latest news from breeders and customers.

There are large annual variations in seed production both in the forest and in seed orchards. In a seed orchard the reproductive output is variable among genotypes and the reproductive phenology may make clones incompatible, so the actual genetic set up of the progeny is a bit uncertain. These uncertainties can be avoided by clonal forestry. Many of our forest trees - like birch - can both spread pollen and produce seeds. That means that selfing can occur and the occurrence of selfing in a plant material will decrease the forest production. With the use of clonal forestry selfing can be completely eliminated.

In a wind pollinated seed orchard wild uncontrolled and unimproved pollen may contribute to pollination, clonal forestry is a way to eliminate this problem. However, for birch, seed orchards are small enough to be managed in plastic green houses, where the inflow of unimproved pollen is low, so for birch it is not fair to see this pollen contamination problem as an argument against open pollinated seed orchards.

In a seed orchard it is important to avoid related clones, as relatives get inbred progeny when they mate, which means that forest production drops because of inbreeding depression. Mating between relatives may be a worse problem than selfing in future seed orchards, as self-zygotes usually die; self-seedlings are often culled before planting; or early out-competed in forest, while milder inbreeding is more directly transferred to production loss. If clones are related in a clonal forest, that does not cause inbreeding depression, and thus there is little harm if related clones occur in a stand (actually this is very natural, trees in a natural stands are often related). The clones in the breeding population will be increasingly more related as breeding goes on, it will be easier to handle the consequences of this for the production forest with clonal forestry.

A seed orchard is a big investment intended to meet the seed need for a long period for a specific species used in a specified area, thus a considerable local market is needed to justify the investment. Conifer seed orchards are usually designed for a need of many millions plants per year over some decades. Multiplication of a clone can be seen as a smaller investment during a shorter period, thus the local market is allowed to be much smaller with clonal forestry, e.g. micro propagation for curly birch in Finland serves a market of 200 000 seedlings a year. A seed orchard has a stiff structure, its composition can be only slightly modified once it is established and it is not easy to justify several orchards heading for different characteristics for the same area. Clonal propagation is more flexible; an individual customer can choose the clones, which fits the particular needs for the customer and occasion. Sometimes a combination of rather rare characters is desired, a clone can be found which combines several desirable characters in a much more efficient way than dealing with seed orchards. It is often stated that clonal forestry reduces genetic diversity. This is a generalisation, which need not always be true. Actually, clonal forestry provides a tool to choose genetic diversity at will, while when a seedling material is used, the genetic diversity is essentially left to chance. Probably those using intensively managed birch plantation for producing high quality timber with special characteristics would choose the option to reduce or eliminate genetic diversity on the stand level, because they found it economically sound to do that. It seems likely this is how the tool will be used. But clones in a clonal mix can be chosen so that the genetic diversity is larger than in a seedling material. Clones can be mixed with seedlings, that may be a good strategy not only for diversity, but economically in cases where cloned plants are much better, but also much more expensive. If the clones succeed well, their share at the final harvest will be larger than at establishment. Over time and space different clones will be used even in a system with monoclone cultures (like Eucalypts in Brazil). In such a system the genetic diversity among stands is likely to be much than in nature. Sometimes clones have been used for natural conservation in situation when good seeds could not be obtained. Clonal forestry may strengthen the trend to produce the needed wood at a limited acreage of intensively managed production forests, allowing larger areas to be set aside for other purposes. Thus the question how clones may affect diversity has many aspects and it is an oversimplification to state that it must be severely reduced.

A common situation is that some good seeds exists or can be obtained, but quantitatively insufficient for the plant need. If such a situation occurs unexpected and unforeseen, vegetative propagation may be the solution. A more common situation is that a few superior seeds can be produced by controlled crosses, but too few to meet the plant demand. Clonal propagation is often a way to make forestry with controlled crosses practical feasible. Clonal propagation may be seen as a way to amplify the impact of a few good parents or a tested cross by mating the parents and when multiply the seeds.

The genetic value of a genotype depends on the breeding value of its parents, but also how well the genes of the parents fit together. This may be called dominance. The later reason for genetic differences cannot be exploited in an ordinary seed orchard, but it can be exploited with clonal forestry.

Hybrids sometimes result in good seeds. It may be difficult to make many of those seeds by controlled crosses, but hybrid seeds (or trees) can be multiplied by vegetative propagation. This is used in the Eucalypts and Aspen programs mentioned above. Personally I am not that convinced that hybrid superiority is so common as it is often stated. When hybrids are better than the parents it may be because the parents were not well adapted for the conditions where they were tested or that the hybrid combines characters of the parents for the environment and purpose of the forester in a better way than tested alternatives. Hybrid maize is a best seller, but the reason is more that farmers are unable to produce hybrid seeds them self, but has to buy them, and when it is good economy of a seed company to invest in hybrids, so they can market the seeds.

To get a genetically superior forest, it is desirable to use the genes of the very best clones. In a seed orchard it has to be rather many parents to avoid selfing and problems with lack of overlap in flowering time. This means a drop in breeding value. For clonal forestry the very best clone can be used. This is true even if in the complete absence of dominance. As dominance is seldom the major cause of genetic variation for the most important characters in forest tree breeding, this more effective exploitation of breeding value is usually a more important reason for clonal forestry than the exploitation of dominance variation, at least as long as we do not talk about hybrids of different species or races. Quantitatively the low importance for the superiority of cloning of a limited amount of dominance variation is demonstrated by e.g. Lstiburek (2000).

We have also special features for clonal selection. We can intentionally select clones, which do not waist resources on sexual activities but spend resources on stem wood production. Such clones are desirable for forestry. In a seed orchard it is not desirable with clones, which do not produce seed and pollen, so such clones would not be chosen for a seed orchard, and if they were, they would be unable to transmit their genes to the seed crop. Cloning methods have different characteristics. They are often very depending on the state of the material which is cloned, cuttings usually do not work well on physiologically old material, for somatic embryogenesis other tissue than immature seeds are often a difficult starting material. With micro propagation it is usually possible to multiply old trees or clones, and this is often an advantage.

Cloning and the cloning method can have effects on the physiological status of the tree. E.g. conifer grafts have not thick bark at the bottom of the trunk. The protocol for making clones can have effects on the plants. These effects sometimes look desirable and can be exploited. For example Norway spruce cuttings seems to suffer a lower mortality from pine weevil damage than seedlings (Mattson and Thorsén 1992). Clones usually origin from older and more mature tissues than seedlings and this affects the characters of the plants and may have effects also on the mature trees. I guess these differences to seedlings are not dangerous and nursery raised seedlings can also be claimed to be unnatural, and they my be different because of different growing regimes. I suggest that we can intentionally interfere with the characters of the plants by the production system and that clonal forestry gives additional options to use this in a positive way. Many of the factors I have mentioned above will contribute to that clonal forestry gives a genetically more improved forest than comparable seedlings. An example of the development of gain over time for the clonal option versus the seed orchard option is shown in Figure 1. This is based on a simulation of the Swedish Norway spruce (Picea abies K.) long term breeding program with inputs chosen to correspond to the real program as close as possible. Clonal forestry and seed orchards are different ways of harvesting gain from the breeding population and clonal forestry means that the genetic improvement reaches a certain level three decades before the same level can be achieved by seed orchards.

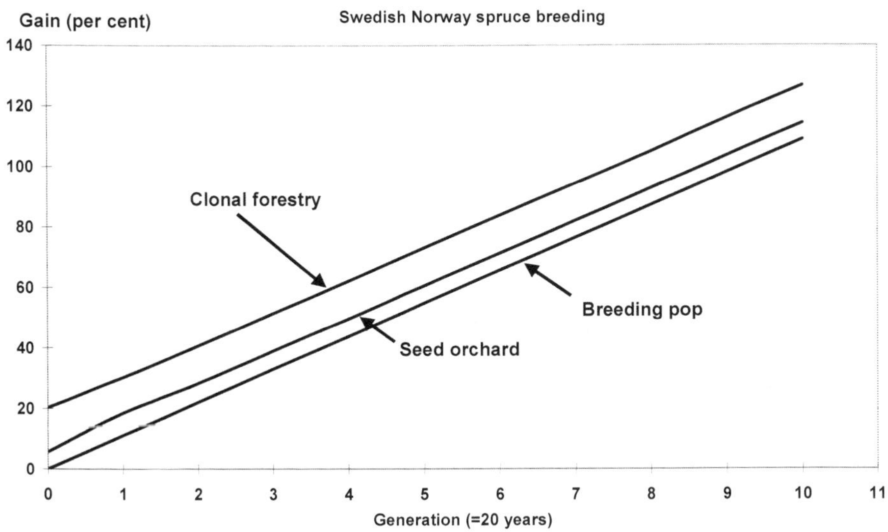

A comparison of the genetic value of three populations as a function of time based on a simulation of the Swedish Norway spruce breeding program. Genetic progress is primarily created in the breeding population. The production population is created by creaming the best available material from the breeding population either by clonal forestry or seed orchards. The gain unit can roughly be interpreted as percent gain. The details of the figure depend on exact specifications, but it gives an idea of magnitudes based on best possible estimates. Clonal forestry gives a higher gain than seed orchards; the difference at a given time is in the magnitude of 15%. The same gain is achievable from seed orchards; it just takes a few decades longer to get it. Modified from Rosvall, Lindgren and Mullin 1998.

More effective breeding with clones!

In the section above it was pointed out that clonal forestry might be a faster way to better forests than seed orchards. Good genotypes are multiplied as grafts in most current forest seed orchards, thus clones are useful also for producing genetically improved seedlings. But there is another - still more powerful - way of using the clonal tool to improve the genetic quality of forests. Clone testing has the potential to make the long-term breeding considerable more efficient. Even if clones are too expensive for direct use in forestry, their use in testing of genetic materials may help forestry to get better seeds in the future. A combination of testing clones for forestry and for long-term breeding has synergistic effects, thus such a program is more efficient than a program focusing only on one of these two aspects.

Higher organisms are diploid, thus have two set of homologous chromosomes and carry two packages of genes. The parents only transfer one package (a haploid gamete) to the progeny, and the composition of that gamete is different for each progeny. Thus, if offspring is used to test parents, only half of the genes in the offspring are relevant for the parent, and these genes will be different for each offspring individual. In contrast, in clonal testing all genes in the genotype are tested and no genes generate random noise in the statistical sense. The performance of a clone depends on its father, its mother and how the genes of the mother and father match together, thus dominance effects. However, dominance variation does not seem large enough for most characters of practical interest to be important (e.g. Rosvall 1999, Lstiburek 2000). Thus the performance of a clone in a clone test says much about what will be inherited to its progeny.

The accuracy by which the breeding value of a clone is determined increases with the number of ramets. The uncertainty caused by the environment decreases then environment is repeated. Thus clonal testing is a way of increasing the accuracy. One can see it as a way of increasing the heritability to decrease the environmental variance by letting the environment be the average of a number of individual environments, one for each ramet. The gain obtained by a genetic selection depends on the selection intensity and the accuracy of the test. Under the constraint of a fixed plant number, a lower number of genotypes can be tested if genotypes are replicated by cloning, and thus the selection intensity will decrease It thus becomes an optimization problem. However, if the individual heritability is not very high and the resource budget is not extremely tight, the optimum is to clone in at least some copies.

A seedling material is genetically diverse; the best genotypes grown pure are much better than the average of a seedling material. The performance of a seedling is mainly depending on its environment and not so much of its genes. If it is cloned, the average over the clone is more depending on the genes. Testing by clones can be seen as a way to eliminate the statistical noise. The heritability of clonal values will be much higher than the heritability of seedling values.

Each seedling is genetically unique; a scientific principle is to use replications when testing. Thus a seedling is a rather unsatisfactory material for genetic testing and conclusions based on seedlings will be dependent on model assumptions, which are uncertain and affected by statistical noise. Clone testing is less depending on model assumptions and a more direct measure and thus offers theoretical advantages!

In a series of investigations (e.g. Rosvall 1999; Danusevicius and Lindgren 2002; assuming conditions similar to the Swedish Norway spruce program), it has turned out favorable to test the breeding value of the candidates for the breeding population by clonal testing instead of phenotypic testing or progeny testing, if that is at all possible.

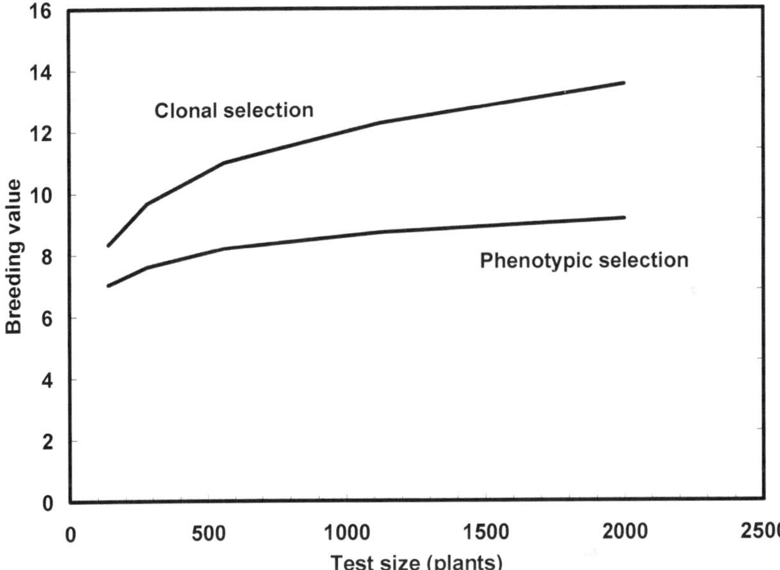

Comparison of clonal or seedling based testing for the Swedish Norway spruce long term breeding program (the X-axes gives the resource level, the test size for the real program is around 560 plants/parent). Clonal testing adds around 30% to the gain of the breeding program, the advantage is larger the more resources that are available for each parent (Rosvall 1999).

The superiority of clonal testing over phenotypic testing is demonstrated in Figure 2 (from Rosvall 1999 Figure 3). It is based on the same type of simulations of the Swedish Norway spruce long term breeding program as the previous figure, although it varies the resources available (thus the size of the tests). The resources available today correspond to 560 test plants per parent. The more resources available, the more reason to use clone testing to get a more accurate selection (compared to increasing the number of test genotypes for a more intense selection).

Lstiburek (2000) and Danusevicius and Lindgren (2002) studied the conditions under which clonal testing is a favourable alternative. The results show that clone test is superior to phenotype test and also progeny test under a wide range of conditions. Thus even if the genotypes or ramets are expensive; and over a wide range of available resources; and in the presence of typical dominance variation; clonal testing can be recommended. In cases where the individual heritability is close to 1, clonal testing is less favourable. When the phenotype itself gives sufficient information about the genotype, the accuracy need not be improved by cloning. Such cases are rare for growth characters, but may be more common for some quality characters.

Numerous studies have shown that environment can cause lasting changes in phenotype that can be passed from one generation to the next, much as genes are transmitted (Schwaegerle, McIintyre and Swingley 2000). Such affects may limit the usefulness of clonal testing for breeding value, but similar objections could be made against progeny testing, and the effects must anyway be larger than usually observed (e.g. Lindgren and Wei 1994) to really matter. Natural genotypes are not designed for perfect cooperation; the Darwinian selection in the struggle for life unavoidable gives some preference to strong competitors. A certain degree of selfishness is an unavoidable consequence of natural selection. By clonal selection we ought to be able to identify and utilise the best collaborators, who act in the interest of the stand and not only the individual tree. Here is a case when Man may have the means to do better than Nature and identify collaborative clones.

A forest material is used over varying conditions in time and space and the forester wants a reasonable stable material over these variable conditions. Clones can be tested over time and space. That make breeders able to select clones, which perform well everywhere and anytime. Natural selection works just here and now. Again a case where Man may find more effective ways than Nature has used. Tree Breeders first select based on field tests and when mate the selections. Artificial crosses are difficult to organise in the remote places where field tests are localised. But if clones are tested, some copies could be placed where it is convenient to make artificial mating, like in archives outside the institute.

Foresters do not want trees to waist resources on sex, but to do breeding it is necessary to get reproduction to work. But clones, which have little reproductive output in the field, can have copies at the research station, which are manipulated to make mating possible. Thus with clones different individuals can be used for test and reproduction. Genetically modified trees will play a role in practical forestry and tree breeding at some time in the future. Such implementation is much easier, if clonal forestry is an established praxis.

EVOLUTION OF TECHNOLOGY FOR CLONING EUCALYPTUS IN LARGE SCALE

Since the first developmental stage until its recognition as an operational method of reproducing superior trees, cloning of Eucalyptus species is undergoing a continuous process of improvement through incorporation of new technical concepts and technologies at different phases of the process.

Campinhos & Ikemori, 1983; Zobel & Ikemori, 1983; Ferreira & Santos, 1997; Denison & Keitzka, 1993.

In the last two decades, cloning Eucalyptus spp. has produced relevant progress for the forest companies, especially solving problems associated with diseases such as canker (Cryphonectria cubensis) and productivity improvement (Campinhos & Ikemori, 1983; Ferreira & Santos, 1997). Actually, the focus of cloning has shifted to industrial requirements, rather than remain limited to just disease resistance and volume increase. The wood properties that positively influence industrial processes and product quality are considered, especially where cloning has an important role to play.

In the era of global markets, the development of forestry plantations for industrial purposes must aim for, besides other objectives, increasing industrial competitiveness in the distinct market segments they interact with. In such scenario forestry based companies must consider the mode in which the forestry raw material can affect their competitive capacity. The modern concept of competitiveness includes producing products to meet the customer's requirements at low costs, in a sustainable manner and with minimum impact on the environment. Therefore, developing tree breeding programs to obtain quick gains and developing cloning systems to have a well established vegetative propagation method becomes important. The vegetative propagation methods should rapidly transform the genetic gains, obtained through breeding, into benefits for the industry. One of the most efficient tools to acquire these objectives is the combination of inter-specific hybridization and establishment of clonal forestry derived from superior hybrid individuals (Assis 2000).

In this context hybridization is an alternative of great impact in tree breeding programs, which can combine superior wood characteristics with tolerance to biotic and abiotic stress, thus representing a significant source of superior individuals, capable of introducing genetic gains in forest productivity and wood properties. Crossing species of different characteristics allows production of complementary wood properties in trees especially to meet industrial requirements.

Considering that Eucalyptus hybrids are heterogeneous, the effective and quick integration of genetic gains obtained with hybrids into the industrial process depends basically on existence of functional large scale cloning systems. Mass vegetative propagation perfectly complements hybridization for producing clonal forestry, and has some advantages over the sexual methods of mass reproduction of selected families, besides being the best way to commercially exploit the heterosis found in several Eucalyptus hybrid crosses. By capturing the total genetic variance (Zobel 1992), vegetative propagation allows for maximum benefits of wood properties and productivity, besides allowing for production of more uniform raw material, which from industrial point of view is highly beneficial to industrial process and product quality. Therefore, the tree breeding programs that focus on these aspects of forest industry will have great impact on the three important components of the competitive process: productivity, product quality and production costs. This article basically describes the recent evolution of cloning Eucalyptus for industrial purposes in Brazil, with emphasis on the cloning techniques and on the systems used for mass production of plant propagules.

NEW CLONING TECHNIQUES

Considering the consolidation of cloning as a tool for the establishment of productive Eucalyptus clonal forestry and its positive effects on industrial process and product quality, the evolution of the systems used for cloning Eucalyptus certainly could be predicted. The precursor of this development started in the early 80's when cloning Eucalyptus by rooting stem-cuttings reached industrial dimension (Campinhos & Ikemori, 1983). Since that time, the major constraint for general adoption was its applicability only to a small number of species and clones. Rooting stem-cutting was not suitable for a large number of economically important species, including those important for energy and charcoal, like E. citriodora, E. maculata, E. paniculata and E. cloeziana, and an important number of clones of rootable species had problems for commercial propagation. Most of the problems were associated with accelerated maturation process causing rapid loss of rooting-predisposition, and manifestation of topophytic effect. The phenomena of topophysis affects clones in different intensities, and is the main cause of intra-clone differences in growth and reduction of rooting ability. Franclet et al. (1987) emphatically pointed out that topophysis induced physiological differences and these differences can result in sufficient intra-clonal variability that can suppress the potential advantages of cloning. Another limitation of stem cutting was associated with alterations of root system architecture, leading to root deformation. In many clones such deformations prevented their full genetic expression, consequently reducing the ratio between selected trees and number of clones effectively used. Because such limitations of rooting stem-cutting, alternative methods were developed for commercial cloning of Eucalyptus species.

Micro-Cutting

Based on the work of Assis et al. (1992), this technique was developed in Brazil in the early 90's. The idea came up with the observations that rooting ability of stem-cuttings decreases with ontogenetic aging and the decline may be faster than reported in the literature. In E. grandis for example, the rooting competence decreased from the fourteenth node up (Patton & Willing, 1974), while it took longer in the E. deglupta. Assis et al. (1992) observed that clones of E. saligna, E. grandis and E. urophylla that had equally high proportion of stem-cutting rooting in vitro, showed differential levels of decline in the rooting percentage when managed in clonal hedges. This indicated that some factor related to clone growth, encompassing period between planting and cutting harvest (6 months), could be responsible for these differences. Preliminary tests done at Klabin Riocell (unpublished) showed, independent the species, almost 100% rooting of the very young mini-cuttings obtained from the cotyledonary leaf pair and the same tendency was maintained in the difficult to root species like E. citriodora, E. cloeziana, E. paniculata, E. dunnii, and E. globulus. However, with age, ranging from few days to some months, the cuttings harvested from such young plants showed a marked reduction in their rooting ability and in some cases such ability was totally lost. These observations suggested that the rooting potential reaches the maximum value at high juvenility level (mini-cuttings from cotyledons) and is similar in all species tested. But the decrease in the rooting ability with seedling age differed among species, which was similar to that found in the older materials in the field. This suggests that, at some stage, part of the juvenility obtained through rejuvenation process in vitro (Gonçalves et al. 1986) and/or on basal sprouts of cut adult tress (Hartney, 1980) is being gradually eroded during the growth of the clones in the clonal hedges.

Thus, hypothetically, the rooting ability of Eucalyptus clones ex vitro should increase if the "physiological distance" between the maximum juvenility stage, obtained in vitro, and the propagule collecting stage is reduced. To test this hypothesis Assis et al (1992) used shoot apex (micro-cuttings) of very juvenile micro-propagated plants of E. saligna clone as propagules. These micro-cuttings had 30% higher rooting than the stem-cuttings. In order to verify the results in a more representative sample, the rooting indices obtained by the traditional and the new micro-cutting method, in seven clones of E. saligna and five clones of E. grandis were compared. Considering all the clones, the rooting of micro-cuttings, on an average, was 18% higher than the rooting of the stem-cuttings, which amounted to an increase in the range of 6.3 to 44.6 percent points over the average. In general, higher increases in rooting percentage were observed in clones with lower rooting ability with the stem-cuttings.

Other results were obtained from several other trials established to define substrate, growth substances, environmental conditions for rooting etc. One of the most significant findings of this new technology was complete elimination of the use of growth substances usually required for the rooting of stem-cutting (Assis et al. 1992). These substances did not increase rooting of micro-cuttings, instead in some cases reduced it, indicating that the endogenous auxin concentration in the juvenile tissues was sufficient to promote rooting. Based on these results, a super-intensive system of Eucalyptus propagation ex vitro was established.

The main feature of the technique is the use of juvenile plants or plants rejuvenated in vitro, as source of vegetative propagules. Shoot apices are used as micro-cuttings, which are placed to root in a glasshouse equipped with temperature and humidity control. The actual size of micro-cuttings is about 7 to 8 cm with two to three leaf-pairs. Presence of the shoot apex is important for quality of the root system, because its presence induces taproot-like system. The micro-stumps left after micro-cutting harvest, sprout rapidly producing new micro-propagules, which can be harvested for use within a period of 15 days in the summer and 30 days in the winter. Since its first use, the micro-cutting technique is improving continuously by incorporating new research findings. The evidence of evolution of this technique are well documented in publications of Assis et al. (1992); Xavier & Comerio (1996); Iannelli et al. (1996); Assis (1997); Wendling & Xavier (1999); Higashi (2000), and Campinhos (2000). The participation of dedicated scientists such as Professor Acelino Couto Alfenas (U.F.V –Viçosa-MG) and Professor Antonio Natal Gonçalves (E.S.A.L.Q.- Piracicaba-SP) among others, was of fundamental importance for the establishment of an efficient super-intensive cloning system for Eucalyptus species. The technical contributions, reciprocally exchanged through a pre-competitive development model and intense information exchange, were the bases for the fast evolution of this new concept of cloning Eucalyptus in large scale.

Mini-Cutting

Probably, the first initiative to root mini-cuttings in Eucalyptus was taken in the early 80's, by using shoots obtained from the thinning operation of rooted stem-cuttings (unpublished). The early trials were discouraging due to inconsistent results that varied from highly positive to negative. Later studies showed that such inconsistencies were caused by nutritional deficiencies of the mother plants at the time of mini-cutting collection. A similar system called "cascade propagation", used in France, is described by Chaperon (1987). However, in the 90's, after the consolidation of the micro-cutting as a functional propagation system, that mini-cutting system became commercially viable for Eucalyptus cloning. Higashi et al., (2000) developed a functional and efficient mini-cutting cloning system and several other researchers added significant contributions to the development of this technique (Wendling et al.2000; Xavier & Wendling, 1988).

Mini- and micro-cutting techniques are very similar in concept and operational procedures, differing mainly in the origin of the initial propagules. The micro-cuttings are obtained from shoot apices originating from micro-propagated plants, and the mini-cuttings come from axillary sprouts of plants cloned by stem-cuttings. After rooting of the first shoot, the two techniques are identical. In some clones using mini-cutting, some propagation cycles (serial propagation) are required to obtain the reactivation for acquiring full potential of rooting capacity. In micro-cuttings such propagation cycles comes naturally by monthly in vitro sub-culturing of explants. Micro-propagation is unnecessary for easy to root species because high levels of juvenility can be obtained easily by inducing basal shoots therefore in such cases mini-cutting is technically and economically feasible.

Despite the success of these techniques is believed to be related to maintenance of very juvenile stage (Assis et al. 1992, Xavier & Comerio 1996), new findings suggest that their high rooting potential are also related to the better nutritional status of the mini-cuttings. In general, mini-cuttings or micro-cuttings root better than juvenile stem-cuttings of the same clones produced in clonal hedges in field. Since both materials are juvenile, the mini-and micro-cuttings cultivated in well-balanced nutritive solutions root better as a result of better nutritional status. The rooting superiority of mini- or micro-cuttings may also be related to differential level of lignification in the two groups of propagules. Compared to stem cuttings, these micro- or mini-cuttings can be considered as "herbaceous", and using these techniques many complication associated with lignin formation and its concentration increase in tissues can be avoided.

Although, the two techniques are very similar, but received different names. The two terms mini-cutting and micro-cutting were designated, by convention, to systems of vegetative propagation that have their origin in rooted cuttings and micro-propagated plants, respectively. Thus a system originating from micro-propagated plants is termed as micro-cutting and that based on sprouts from rooted cuttings is called mini-cuttings. Nevertheless, there is a tendency to change this terminology to avoid confusion with the traditional term micro-cutting reserved for rooting of shoots produced in vitro. Therefore a system of vegetative propagation ex vitro based on mini-propagules could be termed as mini-cutting, independent of its origin, including those originating from a micro-propagation system. Therefore the term mini-cutting will be used to designate both techniques on this paper.

Advantages of Mini-Cutting

Compared to the traditional stem-cutting rooting, mini-cutting has many advantages leading to operational, technical, economical, environmental and quality benefits. Operationally, the labor demanded and cost is markedly reduced, due to elimination of labor intensive treatment with growth substances and many other operations required in an extensive outdoor clonal hedge management system. Many field operations like soil preparation, fertilization, irrigation, cultivation, weeding, pest and disease control, sprout transport etc are replaced by intensive activities in smaller indoor areas at much lower costs, where the amount of chemicals used is also drastically reduced.

The rooting ability of mini-cutting is much higher than the stem-cuttings. Although the benefits vary among the species and clones, in general in poor rootters species or clones with stem-cutting better compare with mini-cuttings, which may increase rooting by 40%. The main reasons for such increase are related to higher levels of juvenility and optimal nutritional content of the tissues, which improves the rooting predisposition and speed of root initiation. The rooting speed of mini-cuttings has two other important consequences in a commercial cloning program: the time of the plants indoor is usually reduced to half compared to rooting stem-cutting, thus considerably improve the use of the facilities and reduce the exposures of basal tissue of mini-cuttings to pathogenic fungi. At the initial and more susceptible stage, the faster reaction of mini-cuttings inducing rapid protector callus formation at the basal end provides protection from pathogen resulting in reduced fungicide application.

Mini-cuttings produce better quality root system with a tendency for a taproot-like system in contrast to the predominant lateral root growth habit in the stem-cutting system. Apparently the connection between root and stem tissues in the mini-cuttings is more suitable due to lower lignification of the tissues involved. This kind of root system appears to be more responsive to fertilization and is more resistant to environmental factors and considerably reduces intra-clone variation. As pointed out earlier, the topophysis effects observed in several systems of vegetative propagation are highly harmful to clonal uniformity. In this respect mini-cutting offers the advantage of physiological homogenization of the propagules producing a much more uniform forest.

EVOLUTION OF MASS PRODUCTION OF VEGETATIVE PROPAGULES

When the rooting of stem-cutting was first introduced commercially in Brazil (Campinhos & Ikemori (1983), the vegetative propagules for establishing clonal forests were obtained from field plantations. This management strategy required annually reserving large areas, with the only objective of obtaining cuttings for commercial rooting programs. Although the commercial plantations could be used to produce propagules, the high time-demand to gather wood and the vegetative propagules made this alternative unfeasible. The clonal hedge based concept of intensive management model was introduced sometime later. This system demanded smaller area due to high shoot productivity per unit area (Carvalho, et al.1991; Higashi 2000). Despite being a tremendous advance for mass production of shoots, clonal hedges were still complicated to manage and poorly controlled. To meet the demand of clonal forestry, the system still required large areas, intensive labor, and large amount of nutrients and water. The propagators were mostly dependent on the weather since the system was completely at the mercy of environmental conditions.

Mini Clonal Hedges

At the beginning of the last decade, the development of micro-cutting technology for Eucalyptus (Assis et al. 1992) materialized the concept of super-intensive management of producing vegetative propagules at commercial scale. Like for cloning methods, the advent of the micro-cutting substantially contributed to the progress made in systems for large scale production of vegetative propagule ex vitro. Originally the system was based on mini-hedges established through rooted mini-cuttings, grown in small containers (dibble tubes). This system provided a series of technical and economical benefits as well as good root quality (Assis et al. 1992; Assis, 1997; Xavier & Comério, 1986). Despite great advance over the clonal hedges in field, mini-hedges faced some limitations. The outdoor mini-hedges were still hostages of climate, and problems related to adequate maintenance of nutritional status and leaf diseases continued, especially during winter. The main problems were, reduced photosynthesis rate, reduced nutrient uptake and high levels of nutrient loss by leaching during period of excessive rainfall or even during irrigation. These limitation led to the development of indoor mini hedge system.

Indoor Hydroponics Clonal Hedges

The development of new technologies contributed to increase the efficiency of the existing systems. Hydroponics concepts for production of mini-cuttings was introduced first in operational indoor system based on drip fertigated sand beds (Higashi et al. 2000). The major contribution of this system was in terms of assuring an adequate nutritional status of the mini-cuttings, which is the key factor to obtain high rooting percentage in juvenile vegetative material. Using the same concept, Campinhos et al. (2000) developed a highly efficient method based on intermittent flooding system, where the containers of the mini-stumps are immersed in a nutritive solution for fertigation. These two hydroponics methods, sand-bed and intermittent flooding, are most widely used and the Brazilian companies use either of them.

Virtual Clonal Hedge

The observation that the shoots of first collection root better than those after third or fourth collection, suggested that vegetative propagules could be produced anywhere, without allocation of a definite physical area. Assis (1997) suggested a system where the mini-cuttings can be produced in virtual clonal hedges. Any rooted plant with normal growth when reaches a height of about 15cm, has its apex (7 to 8 cm) cut back to produce a new mini-cutting. As soon as the newly rooted mini-cuttings are about 15 cm high, their apices can be used as a new source of mini-cuttings, thus creating a virtual mini-hedge after some cycles of re-rooting, and requiring no specific area for production of vegetative propagules. The important attribute of this system is that the plants used for producing mini-cutting can be planted in field without losing quality of the intermittent flooding mini hedges.

The sprouting of harvested plants becomes an extra option of providing mini-cuttings. The blanks resulting from rooting failure can be filled with mini-cuttings produced on the mini-stumps of the harvested plants. Similarly, the sprouts produced on the pruned plants and the apices of the rooted mini-cuttings can be used continuously as self-sustained cyclic process. This system represents an excellent technical, economical, and operational alternative, but only works well if good nutritional conditions of the mini-cuttings are ensured. An extension of sub-fertigation to newly rooted mini-cuttings is the most suitable option to make virtual mini-hedging practical. The use of very soft mini-cuttings must be avoided to prevent fungal infection. The locations with wide climatic variations and without any environmental control are less suitable for establishing virtual mini-hedge system, therefore is not recommended in situations where environmental control is not feasible.

Advantages of Indoor Hydroponic Mini Hedging

The development of the concept of super-intensive hydroponic mini hedging, brought several benefits to commercial cloning of Eucalyptus. All indoor mini hedging systems have technical and economical advantages relative to normal outdoor clonal hedging. Higher productivity of mini-cuttings, lower labor demand, and low consumption of chemicals and water, represents major sources of economic gain. In addition, considering the important effect of the conditions under which the mother-plants grow, on the rooting success (Read, 1987; Hartman & Kester, 1983; Haissig, 1986; Andersen, 1986), the main technical advantages are related to such aspect. These systems allow the use of important theoretical concepts which are difficult to implement outdoor, such as CO2 enrichment, control of temperature, light intensity and photoperiod. These factors in addition to the nutrition are of fundamental importance to enhance rooting predisposition. This system also enables application of pre-harvest treatments such as spraying mother plants with growth substance (phyto hormone) to increase rooting potential.

Propagules Productivity

Eucalyptus cloning in large scale started in Brazil using very extensive systems and the productivity of vegetative propagules was 114 cuttings/m2 of hedges field(Campinhos & Ikemori, 1983). With the introduction of the mini-hedges the production increased progressively to 121 cuttings/m2 (Carvalho et al. 1991), then to 1,752 cuttings/m2 per year (Higashi et al. 2000) and presently is at about 24,000 cuttings/m2/year. Table 1 provides an overview of the evolution of vegetative propagules productivity in Eucalyptus. The super-intensive systems, especially hydroponics based, are much more efficient, having productivity 350 folds higher than the initial systems used in Brazil. Comparing the different systems, the capacity of producing vegetative propagules in the super-intensive systems is highly expressive, consequently, more cuttings can be produced per year per unit constructed area of mini hedge, besides they are ease of handle, manage and they have a lower cost per plant.

Variations in the productivity of mini-cutting observed in table 1 resulted from different hedging systems, from different genetic materials and also from different management practices. Clones of E. grandis x E. urophylla hybrids are generally much more productive in mini-cuttings than other species. The majority of the pure E. saligna clones are much more difficult to produce as mini-cuttings than the E. grandis x E. urophylla hybrid where production/ mini-stump is almost double. Still, the rooting competence and propagule productivity is much higher than in the outdoor clonal hedges. The other source of variation in the productivity is related to the concept and kind of management adopted to utilize the sprouts.

Certain practices of management of mini-stumps in the sand beds produce vegetative propagules similar to the macro-cutting system. The sprouts are allowed to grow to bigger sizes, and after collection each sprout is divided into three mini-cuttings (basal, middle and apical). This system produces higher number of propagules/m2 but it loses the best characteristic of the mini-cutting provided by the presence of shoot apex, responsible for producing high quality root system (taproot-like system, discussed earlier)

In both hydroponics systems (drip fertigation and intermittent flooding) losing mother plants is common due to salination, excessive harvesting, diseases etc. In this respect, the systems that use containerized mother plants, as in intermittent flooding, have advantage over the sand bed. In sand beds, the dead plant replacement and getting them to productive stage require more time compared to the intermittent flooding system.

The most attractive characteristic of virtual mini-hedges, that do not require allocated physical area for producing vegetative propagules, is that the mother plants are plantable. After supplying mini-cuttings, such plants are managed as normal rooted mini-cuttings. In the intermittent flooding system, the mother plants also can be planted provided the harvesting regime has not degraded its quality. Under successive and heavy harvesting regime the quality of mother plants frequently declines.

One question that arises from the spacing in mini-hedges is about the influence of close spacing on the rooting of mini-cuttings and the mini-stump productivity. Trials at Klabin Riocell, showed that the spacing of 5 x 5 cm between plants does not interfere significantly in the rooting capacity of the mini-cuttings, but wider spacing permits higher survival rate of mini-stumps. But the final results are better with narrower spacing, since there is more than proportional compensation by existence of higher number of plants/m2.

Table. Propagules productivity in different clonal hedging systems used in Brazil*.

System	Species/ Spacing (m)	Productivity (Propagules/m²/year)	References
Clonal bank	E. grandis x E. urophylla - 3,0 x 3,0	114	Campinhos & Ikemori (1983)
Clonal Hedging	E. grandis x E. urophylla - 1,0 x 1,5	121	Carvalho et al. (1991)
Clonal Hedging	E. grandis x E. urophylla - 0,5 x 0,5	1752	Higashi et al. (2000)
Mini hedging Out door	E. grandis x E. urophylla - 0,05 x 0,05	29200	Xavier & Comério (1996)
Hydro mini hedging (sand bed/drop irrigation)	E. grandis x E. urophylla - 0,1 x 0,1	41480**	Higashi et al. (2000)
Hydro mini hedging (interm. flooding)	E. grandis x E. urophylla - 0,05 x 0,05	25500	Campinhos et al. (2000)
Hydro mini hedging (interm. flooding)	E. grandis x E. urophylla - 0,05 x 0,05	24000	Klabin Riocell operational
Hydro mini hedging (intermittent. flooding)	E. saligna and E. urophylla x E. globulus - 0,05 x 0,05	14400	Klabin Riocell operational
Virtual mini hedging	E. saligna and E. urophylla x E. globulus 0,05 x 0,05	-	Klabin Riocell operational

*Based on cuttings/mini-cuttings/stump considering 12 months of effective production, without discounting the replacement of dead plants.
**Subdividing sprouts in 3 mini-cuttings

Temperature and Light

Temperature and light are good examples of the factors that can be controlled in mini hedges, to improve rooting predisposition of mini-cuttings. Light can be managed as to photoperiod and intensity. Light intensity can strongly influence cutting productivity and rooting by reducing or increasing endogenous phenolic substances that act as rooting inhibitors (Vieitez & Ballester, 1988) or promoter depending upon the concentration in the tissues and the species involved. The photoperiod, especially combined with temperature can influence rooting predisposition of shoots. Lighting also can influence rooting by controlling internal levels of carbohydrates. Etiolation of mother plants can affect rooting and root number, which depending upon the specie may increase or decrease (Hansen, 1987). This variation appears to be linked to carbohydrate production (Nanda et al. 1971) and accumulation and transport of auxins (Hansen, 1987).

Temperature can influence rooting by interfering with nutrient uptake and metabolism, and its control, especially in sub-tropical climates, can be adjusted for optimum production of cuttings. Assis (1997) observed that rooting percentage of mini-cuttings decreases in cold winter months. Recent work at Klabin Riocell (Unpublished) showed that this problem could be solved by supplying additional light (14h/1000 lux) and increasing the temperature in environment of mother plants (>20 oC). Since the two factors were not separated, the effect of individual factor could not be determined. However, considering that nutrient concentration is of prime importance for rooting process, and that their uptake depends on the metabolic activity, it can be assumed that both the factors contributed to reestablishment of normal rooting competence. In general, in areas of cold climate, rooting is limited to only warmer months, therefore as discussed above, if light and temperature are controlled the use of greenhouse space can be optimized by extending the rooting period for the entire year.

Nutrition

The most important advantage of hydroponics mini-hedges is the possibility of providing well-balanced nutrition to mini-cuttings. The effect of macro- and micro-nutrients on root initiation of cuttings on several plant species, has been well documented in the literature (Andersen, 1986; Read, 1980; Hansen, 1987; Haissig, 1986; Hartman & Kester, 1983). As discussed earlier, nutrition is the key factor affecting rooting predisposition, because of its involvement in determining the morphogenetic response of the plants. In hydroponics mini-hedging system the nutrients are supplied in ideal concentrations to promote rooting. Hydroponics, without the influence of rainfall and or overhead irrigation, allows better nutrition control of mother plants. The use of indoor hydroponics mini-hedges allows nutrient supply in a correct concentration, avoiding the nutritional imbalance commonly observed in outdoors clonal hedges; besides there is no leaching by rainfall and the response to nutrient application is rapid. According to Haissig (1996), a feasible nutritional balance of mother plants can be linked to production of tryptophane, the precursor of IAA and to storage substances.

Calcium plays an important role in the rooting process, mainly because it acts as peroxidase activator (Haissig, 1986). According to practical experience of Klabin Riocell, the percentage rooting of mini-cuttings decreased considerably if the leaf calcium concentration dropped below 0.7%. This situation was observed especially in winter when, in the absence of heating, leaf calcium concentration was substantially lower than in warmer months. Calcium deficiency induces shoot apex necrosis in mini-cuttings, which can cause problems of rooting and mini-cutting development (McComb & Sellmer, 1987). This phenomena was commonly observed on the initial, operational scale, outdoor mini hedges.

In general, moderate deficiency of nitrogen can improve rooting (Haissig, 1986), but at high levels more energy is required for vegetative growth and especially for leaf expansion. Consequently, carbohydrates are not stored at suitable levels reducing the C:N. On the other hand, extreme nitrogen deficiency can reduce rooting since it is necessary for nucleic acid and protein synthesis. Based on the considerations of Hartman & Kester (1983), and Haissig

(1973b, 1986) a general model for leaf nutrient concentration can be established. The mother plant must be well nourished with macro nutrients like phosphorus, potassium, calcium and magnesium and moderately deficient in nitrogen.

The effects of micro-nutrients on rooting are variable. Manganese, due to its inhibitory effect should be used at minimal concentrations, but boron and zinc are essential for rooting process (Andersen, 1986). Zinc increases the endogenous content of auxins by increasing trpytophane proportions while boron is involved on rooting and root growth. Table shows nutrient concentrations in leaves of well nourished plants (Roberto Ferreira Novais, pers. Comm.) for rooting mini-cuttings. Monitoring leaf nutritional concentration of the mother plants, as a guide to assure these levels of macro and micro-nutrients has been an important factor in improving the rooting competence of mini-cuttings.

Table. Concentrations of macro- and micro-nutrients in leaves of well nourished Eucalyptus plants.

MACRO NUTRIENTS (%)					MICRO NUTRIENTS (mg/kg)					
N	P	K	Ca	Mg	B	Mn	Zn	Fe	cu	S
2,5 to 3.0	0,2 to 0,4%	1,5 to 2,0	1,0 to 1,5	0,25 to 0,40	40 to 70	100 to 500	50 to 60	100 to 200	10 to 15	0,15 to 0,25

Phytosanitary Advantages

Shoot contamination by splashing of soil during rainfall or over-head irrigation (Ferreira, 1996) is a very well known sanitary problem associated with outdoor clonal hedges. As many pathogenic fungi are soilborne, the cuttings produced either in clonal banks or in outdoor clonal hedges, are frequently contaminated by these fungi. Thus one of the advantage of using indoor hydroponics is the preventive phytosanitary control. In the absence of leaf wetness occurrence of leaf diseases is greatly reduced, thus producing pathogen free cuttings that increases the results of rooting. However, all these benefits of indoor system can be lost or become disadvantageous in the absence of efficient controls to minimize effects of external factors such as extreme temperature and humidity.

MANAGEMENT OF MINI-STUMPS IN HYDROPONICS SYSTEMS

The management of mini-stumps is rather simple, but some aspects must be considered for the technique to be successful. Mini-cutting collection should be done selectively to avoid mini-stump degradation. Desalination should be done at least once a week by generously irrigating the substrate to leach the surface-accumulated salt. In non-automated nutritive solution preparations, the entire solution should be changed every 15 days. In intermittent flooding better results are obtained when the electric conductivity is adjusted to 1,8 –

2,2 in the winter, and to 0,8 – 1,0 in the summer. Avoiding cut and abscised leaves to fall into the nutritive solution (Alfenas, 2001) prevents biotic and abiotic problems on mother plants.

USE OF MINI-CUTTING FOR CLONING OTHER WOODY SPECIES

The applicability of mini-cutting technique has been tested also on other, broad leafed and conifers woody species. In general, the methods used are similar to those of Eucalyptus, with minor adaptations. The results have shown that this technique can be fully applied to species like Pinus taeda, P. elliiottii, Acacia mearnsii and Ilex paraguaiensis and many other woody and non-woody species. The best results were obtained with juvenile plants as is the case with difficult to root E. citriodora, E. maculata and E. paniculata. Thus, mini-cutting is feasible to be used consistently for family multiplication of these species. Use of mini-cuttings can be a possible alternative for production of family forests by multiplication of full-sib or tested half-sib families.

Klabin Riocell started a small program of establishing E. globulus clonal plantation, based on adult and early selected trees resistant to rust caused by Puccinia psidii and other diseases. The preliminary results suggest that mini-cutting is a very promising technique to commercially propagate this species. Compared to stem-cutting, the mini-cutting gave better results in rooting proportion and root system quality. It appears that the propagation of E. globulus is profitable, since it is considered to have different and higher nutritional requirements. The development of special nutritive solutions for E. globulus tends to improve the results and it is clearly possible to establish efficient clonal program for this species.

In Acacia mearnsii, mini-cutting has shown good results, despite being considered as recalcitrant species with regard to vegetative propagation (Matheson 1990). Despite reasonable success obtained by Assis et al. (1993) in adult trees, application of mini-cuttings gave better results, indicating possibility of developing clonal forestry of this genus. The key point in this case is to achieve juvenile or rejuvenated shoots to start the process. Through induction of basal shoots, without tannin oxidation normally present in the chopped trees of this specie, it is necessary to establish the initial healthy rejuvenated plants, which should done using standing trees to avoid problems with tannin.

In Pinus, only juvenile seedlings were tested as mother plants. Pinus elliottii seedlings can be easily rooted by mini-cuttings with more than 85% success. Pinus taeda also roots well though somewhat less (76%) than P. elliottii. The results suggest the possibility of developing forestry families programs with both species based on the systems used for Eucalyptus. Managing Pinus mini-hedges in hydroponics is efficient for providing a good nutritional status to the mini-cuttings. The productivity of mini-propagules, without any change in the nutritive solution, was 9600/m2/year. The results of mini-cutting productivity and rooting proportion are highly encouraging.

10

Genetically Engineered Trees in Forestry

INTRODUCTION

Intent

This paper will address the question of whether there will ever be the certification of forests that contain genetically modified (GM) trees. This chapter, chapter two, reviews the origin and range of both certification systems and genetic modification (GM) applications, and defines a genetically modified organism (GMO). The third chapter will focuses on the aspects of GM that certification schemes, and the Forest Stewardship Council (FSC) in particular, find problematical. It doing so it will consider how the different certification schemes have addressed comparable issues to those raised surrounding GM trees. This includes the conservation of genetic resources, use of exotic species and application of chemicals. Chapter four briefly reviews the inherently different perceptions of GM by the various stakeholders, and the role this plays in risk evaluation. Lastly, chapter five concludes with consideration of the politics and business surrounding GM and certification, and reflects upon their greater role in determining whether forest certification and genetically engineered trees will ever be compatible.

The rise of certification

International responses to deforestation have been vociferously criticised by many NGOs . These reproaches centre on the perception that there has been a slow implementation of inadequate forest conservation agreements. Furthermore, these agreements themselves do not commit participants to agreement implementation. These deficiencies have been recognised for some time; an FAO report on TFAP noted an "inadequate regard for local peoples, ecological matters and the underlying causes of deforestation" and recommended "avoid bureaucratic suffocation and encourage effective leadership" (Ullsten et al. 1990). The ITTO has been consistently criticised as unrepresentative with ineffective environmental concern and poor management, ignoring their prescribed objectives, such as Target 2000 (Colchester 1990; FoE 1992). Moreover, the impasse between 'North and South' has, in the eyes of many NGOs, lead to painfully slow progress since the UNCED 1992 summit; many countries have yet to ratify agreements, and international initiatives (IPF, IFF, ITFF and UNFF) established to further UNCED work, have made slow progress (Humphries 2000).

However, from these processes, SFM emerged as a new framework. It aims to describe management that ensures long-term forest health and productivity, while providing continued social and economic benefits (Evans 1996). In 1988, the ITTO noted that only 0.08% of tropical forests were managed in this manner (Poore 1989). Although endorsed in principle, SFM is not well understood in operational terms. Currently, principles, criteria, and indicators are being developed as a means to assess and report on progress towards SFM. These have been developed largely through international agreements building on UNCED 1992; principally the Helsinki Process, the Montreal Process, the Tarapoto Proposal, a renegotiated ITTA and Dry-zone Africa (Grayson & Maynard 1997).

It has been noted that deforestation continues apace, and legislative measures to curb unsustainable measures have largely failed due to international squabbles, general apathy and agreements that are non-binding (Dunleavy 1993). Further, the de jure appearances of legislation are seen to ignore the de facto realities of implementation at the forest level. A market-based mechanism of certification was considered as a means of circumventing these failings. Being site-specific, certification could validate 'on-the-ground' operations as employing the best management practices through criteria and indicators (Upton & Bass 1995). Consumers, using a labelling system attesting to this SFM, were to discriminate between different production methods. Those companies who were certified could gain many benefits, chiefly improvement / retention of market share, defence against environmental criticism, and investor assurance. This would drive the implementation of SFM in the wood product trade. Those who continued to practice and trade in uncertified products would have difficulty marketing their produce. Credibility was seen as a precondition for any successful certification scheme; independence and third party assessment were understood to be requisite (Upton & Bass 1995).

Certification is a voluntary process, which results in a written statement attesting to the origin of wood raw material, and its status following validation by an independent third party (Ghazali & Simula 1994). Certification typically includes two main components: forest management certification and product certification. Forest management certification is based on an assessment of forest management against a set of standards reflecting contemporary concepts of sustainability. Product certification involves verifying the chain of custody of wood from the certified source to the consumer. This whole process has faced repeated criticism (Counsell 1996; Centero 1998).

Nevertheless, governments and companies are examining certification with interest; it has become a "high profile subject in the forestry sector" (FAO 2000a). Despite the stated purpose of certification for improving forest management, the main interest of most of those undertaking certification at present is probably the marketing benefits it may offer. This may explain why over 80% of FSC-certified forests are in developed countries, and 66% are by industrial enterprises (Thornber 1999). However, in developing countries it is noted "certification serves as an added strength as it facilitates entry into foreign markets" (MTB 1999).

Increasing numbers of certification schemes are being developed; international, national and regional, all can be split into two types.

- The performance based approach, best represented by the FSC and some emerging national schemes (Indonesia, Malaysia and Finland), is founded on standards that an organisation has to meet before it can be certified. This method has been criticised because, although certification is an incentive for forests that are already well managed, companies managing forests poorly may find adjusting their practices very difficult. Consequently they may simply ignore certification (von Maltitz 2000).
- The process-based approach, which is the basis of the system adopted by ISO 14000 series of EMS verification, assesses the quality of an organisation's management process; how it sets its policy and management objectives, and how it organises itself to deliver them consistently. They do not lay down specific pre-determined performance standards, but rather look for continuous improvement of environmental management. The ISO 14000 series, and in particular the forestry-specific 14061, ensure companies set increasingly higher standards and achieve them. This approach is criticised as a means of gaining certification by compliance with internal policy, whilst still being environmentally irresponsible. Moreover, 'ring fencing' can give a misleading impression that ISO standards are applicable to the whole of a company. Lastly, since ISO places a great emphasis on achieving national standards, it could be argued it maintains the disparity in performances between countries, and consequently misleads consumers (von Maltitz 2000).

Technically, ISO certify management systems and not forests, and as such they do not enable product 'labelling'. It is probably for this reason that ISO implementation is greater in companies supplying predominantly to the pulp and paper sector, whilst companies supplying wood timber opt for performance-based systems, such as the FSC. For example, all SAPPI's plantations in South Africa are ISO-certified, but only those supplying saw-logs are FSC-certified (von Maltitz 2000). This contrasts with nearby Mondi plantations, with over 430,000 ha of FSC-certified timber forest producing only saw-logs (FSC 2000e). ISO certification meets pulp and paper consumer demands, and thus serves as an adequate marketing tool. However, ISO has been used by a number of companies as a step towards gaining product certification (Bass & Simula 1999).

Increasingly, process-based schemes are incorporating performance targets; correspondingly, performance-based schemes acknowledge the benefits of EMS (Kanowski et al. 2000). Many see the FSC and ISO approaches to certification working in tandem , complimenting international and national policies to ensure SFM is implemented at all levels. There is little support for the idea of harmonisation of certification systems, but growing support for mutual recognition. For example, six certification systems, including SmartWood (US Rainforest Alliance) and Woodmark (UK Soil Association) are harmonised with the FSC, and their standards are mutually recognised.

Prominent NGOs support the FSC process because they consider its standards to be the most stringent and open too little 'interpretation'. These supporters are concerned that mutual recognition should not diminish FSC standards. However, many advocates of alternative certification initiatives interpret this concern, rightly or wrongly, as an attempt to maintain the dominant role of the FSC in certification. Regionally-based systems are often espoused as better tailored to meet inherently different local circumstances, and commercial plantations perceive the FSC as bias towards natural forest management. However, the FSC has committed itself to collaboration with Malaysian and Indonesian certification initiatives, with the eventual objective that certification, by either party in those countries, might be recognised by both parties (FSC 1999d).

The development of genetic modification (GM)

Paralleling the rising concern over deforestation, much scientific and commercial attention has focused on improving the genetic stock of forest trees in order to improve productivity and quality. The demand for wood-based products is forecast to double over the next 17 years, the onus being on paper and board products (Soil Association 1998), which must be met from plantation-grown trees. GM and other biotechnology techniques have provided conventional tree breeders with new tools to meet this demand, enabling the accelerated modification of some forest trees possible (Dinus & Tuskan 1997). The potential gains are great, since trees remain genetically and phenotypically very similar to their wild progenitors (Wright 1976) and GM can progress more rapidly than conventional breeding, which can take decades.

The first commercially interesting genes available to GM in forestry were those which were developed in agriculture, and thus involved the transfer of DNA between taxonomically distinct organisms. Traits currently being researched include herbicide resistance, increased vigour, pest and pathogen resistance, increased tolerance to biotic stress and improved timber quality (Riemenschneider et al. 1988; Bauer 1997; Dickson & Walker 1997; Tzfira et al. 1998).

Recently, particularly in Europe, GM has been at the centre of a heated debate. This has concentrated on agricultural applications, but has occasionally touched forestry. The issues are enormously complex and not related to purely scientific questions: there are questions of philosophy, ethics, equity, responsibility and ownership. Most proponents have called for "credible information that promotes rational debate" ensuring the public can make an "informed evaluation" (Strauss et al. 2000b). Some of the media have also put forward what appears an assessment of potential benefits and risk (Weiss 2000). However, the media are usually criticised for hyperbole. In the UK, "pseudoscience" and "alarmist media reports" from pro- and anti-GM camps have confused matters and brought GM into the limelight (Nuffield Council on Bioethics 1999). Opinions are sometimes extreme. Some call for deregulation and accelerated introduction of GMOs for economic and environmental reasons (Cantley 1998), whilst many environmental groups are very wary of potential impacts, and consequently believe "genetically modified organisms must not be released into the environment" (Greenpeace 2000).

The UK is an interesting case. The public generally perceive the attitudes of the large companies pursuing GM as irresponsible and principally profit driven. Furthermore, consumer confidence in food is very low following episodes of salmonella in chicken eggs, E. coli and BSE in beef, and the reprocessing of condemned poultry for human consumption. Coupled with extensive media coverage, these two factors have acted synergistically, and the public are sceptical of the motivation behind, and safety surrounding, GM. On 11th July 1999, campaigners destroyed most of the trees in a research trial of GM poplars established by AstraZeneca. This action illustrated how strongly some people felt about the potential ecological impacts of GM. These impacts include gene flow, invasiveness and unpredictability (Owusu 1999). However, there are also potential ecological and economic benefits (Mathews & Campbell 2000).

The advantages and disadvantages of GM trees have been considered by many (reviewed by Mullin & Bertrand 1998; Mathews & Campbell 2000), but as yet no GM trees are in commercial production. Big-business is regarded by many as an outright advocate of GM plants (Allen 1999). Conversely, the FSC and its allied certification schemes, advocating environmental campaigners concerns, have banned outright the use of all GM trees in the forests they certify (FSC 1997, 1999b, 2000a; Soil Association 1998). However, many scientists believe there is a middle-ground between these two views, and have noted "the opponents of the technology have framed the issue as black and white. GMOs are dangerous and must be stopped. Proponents are faced with the difficult task of trying to educate the public about the many shades of grey" (Somerville 2000). This has raised debate about the rationality behind a complete ban on GM trees, for "each GMO should be assessed on its own merits" (McHughen 2000, p.113).

GM THROUGH CERTIFICATIONS EYES

Certification bodies can be placed in two groups; against GMOs and tolerant of GMOs. The first camp is centred solely on the FSC; Principal 6.8 of FSC guidelines states the "use of genetically modified organisms shall be prohibited" (FSC 2000a). This is reiterated in national standards; line 266 of the FSC Forest Management Standard the United Kingdom states simply insists "genetically modified organisms (GMOs) are not used" (FSC 1999b), as does 6.5.11. of the Swedish FSC Standards (FSC 1997). Certification systems recognised as equivalent to the FSC, such as The Soil Associations' Woodmark (section 5.610) also prohibit GMOs.

For many large companies considering certification, the FSC is one of the international systems, largely due to its international scale and backing from influential NGOs such as WWF, Greenpeace and Friends of the Earth. Although the FSC have yet to publish an official document qualifying their policy towards GM, supporting NGOs have disclosed their concerns (Owusu 1999) and some resolutely reject the use of GM plants (Greenpeace 2000). This is important because these NGOs are considered to hold substantial influence over FSC policy . It is likely FSC policy will parallel much of their supporters' concerns, both now and in the future.

The Soil Association published their policies regarding GM, and note that "it would be sensible to apply the precautionary principle and not release GMOs into the environment, at least till their long term impact has been determined" (Soil Association 1998). When discussing certification accreditation they state "this position will be kept under constant review but the evidence that is continuing to emerge confirms the need for prudence and reinforces the logic of adopting a precautionary principle in conjunction with a comprehensive programme of research and monitoring" (Soil Association 1998). Thus, if research and monitoring of GMOs satisfy certifiers' concerns, GM and certification may be compatible, despite the current FSC stance regarding GM.

The other certification camp is more varied. Although it claims not to be certification system, the SFI standard of the AF & PA has all the trappings of, and is considered by others to be, a certification system (Kiekens 2000; McKeand, pers. comm.). This U.S. programme is industry-lead and supported by a number of NGOs, although not nearly with the profile of those that back the FSC. However, the SFI program encompasses more than 26.8 million hectares (Patrick 2000), and by the end of 2001 AF & PA projects that 12.1 million hectares of member company forestlands will have been third-party certified (Pulp & Paper 2000). This compares with just 1.8 million hectares of US forest currently certified by the FSC (FSC 2000b).

What is a GMO?

We lack a precise and common definition of a GMO. The FSC adopted a definition that combined principals outlined in EC Directive 90/220 with the UK Government Health and Safety Executive publication on Contained Use of GMOs (EU 1990, UKGovernment 1996). This indicated a "Genetically modified organism (GMO) means an organism in which the genetic material has been altered in a way that does not occur naturally by mating and/or natural recombination or both " (FSC 1999a). This definition encompasses DNA introduction via living vector systems, such as Agrobacterium tumefaciens used with poplars and eucalyptus (Griffin 1996), and mechanical DNA introduction methods, such as biolistics, used with Pinus radiata (Sederoff & Stomp 1993). Moreover, 'unnatural' cell fusion and hybridisation techniques would be classified as GM. All other technologies were considered permissible.

The continuum and complexity of different breeding technologies makes distinguishing between 'traditional' breeding from genetic modification very difficult. Humankind has been genetically altering crops for millennia (Diamond 1997). This has resulted in varieties almost unrecognisable from their wild progenitors, and these changes may be very rapid. For example, the apical dominance of maize (Zea mays spp. mays) makes today's cultivars strikingly different to their wild progenitor, teosinte (Z. mays spp. parviglumis). This change results largely from alteration of the regulator sequence of the teosinte branched1 (tb1) gene (Doebley et al. 1997). The remainder of the tb1 gene is identical in both maize and teosinte (Wang et al. 1999). Moreover, the distinction between breeding systems may be inconsistent with the FSC's stated concerns over GM (FSC 1999a). For example, herbicide resistant plants have been derived both 'conventionally' and through GM (Concar 1999). The potential risks of gene flow and enhanced weediness from such cultivars are considered to be very similar to GM cultivars. Indeed, some accepted methods may be riskier than GM. For example, the FSC notes specifically that mutagenesis is permitted (FSC 1999a), a process that involves exposing cells to ultraviolet light, radiation or some of the "most poisonous chemicals known to humanity" (McHughen 2000, p.65). The genetic changes this induces are usually not understood. GM has knowledge of exactly which genes have been introduced, and how they function (MacKay et al. 1999). Leading research panels have concluded that the genes and traits must be examined, rather than the method of genetic modification (National Academy of Sciences 1987; National Research Council 2000).

This dichotomy between anti-GM and pro-GM camps reflects an underlying philosophical concept, 'natural' and 'unnatural' , and it is crucial to understanding the debate about biotechnology (Kershen 2000). The label applied to a product depends on the 'view' of the concept of 'natural'. One influence on this view may be the FSC themselves, yet other possible influences include the views of those who lobby the FSC. It must be remembered that certification is a marketing exercise to help promote SFM, and the FSC must create a brand, a label, respected by the consumer. The consumer must believe that certification testifies to sustainable forest management. Thus, the view of 'natural' is imperative. If the FSC feels more akin to environmental groups or more politically influenced by these groups or a 'green' public, the FSC may endorse the 'nature' view with concomitant consequences for regulatory policy; GM will not be certified since it is an 'unnatural' artefact of man's activity.

It is this philosophical divide which is paramount in the debate surrounding genetic modification . GM is viewed by many 'environmentalists' and members of the public as intrinsically different from other advanced breeding techniques, including mutagenesis, embryo rescue, somaclonal variation and cell selection. This may reside in the ability to insert genes from taxonomically distinct species . It is not entirely rational, since many crosses between plants of the same species involve plants grown so geographically dispersed that they would never mingle DNA without human intervention. However, the ecological impact of producing hybids and intoducing exotic species are cited by some as a case for the possible environmental impact of GMOs (Owusu 1999). The need for objective evidence is required by both sides of the debate, and substantive equivalence should be invoked when considering how valid the concerns surrounding GM are.

Even the impact assessment of the risks of GM is fraught with philosophical arguments. As noted by Apel (2000), it is necessary to focus on those factors relevant to an objective risk assessment, based on scientific reasoning. However, it is not clear how this might be accomplished in practice.

THE SPECIFIC CONCERNS OF CERTIFICATION BODIES

The FSC are the only forest certification body to have banned the use of GM trees in the forests that they certify (FSC 2000a). This ban was imposed due to concerns about the environmental safety of GMOs. The concerns have been published in an unofficial document (FSC 1999a), and most are identical to concerns expressed by environmental NGOs (Soil Association 1998; Owusu 1999; Greenpeace 2000). The following chapter focuses on the FSCs stated concerns about GMOs, and considers each issue by discussing equivalent practices examined by the FSC. Where appropriate other certification systems are considered.

Reduced 'diversity'

GM technology is likely to be employed by fine-tuning genera and species already adapted to a site and which can be mass propagated for direct use in plantations (Griffin 1996). This has caused concern within the FSC: "Plantations using one or few transgenic clones will contain less landscape-level diversity than is currently found in plantations using species or varieties resulting from traditional tree-breeding" (FSC 1999a). Yet this misgiving is not restricted to GM per se. Clonal forestry is already well established for Eucalyptus and Pinus radiata, and has been performed with Populus species for over one hundred years (Muhs 1993). The risk of reduced diversity is equally applicable in any clonal plantation. Diversity requirements should be established regardless of whether a tree is GM, clonal or exotic; certification standards recognise this.

PEFC guidelines for Forest Management Practices aim to maintain and 'appropriately' enhance (PEFC 1998). What exactly is meant by genetic diversity is not made clear in this document. Furthermore, these guidelines expounded very little at a national level (PEFC 1999). For example, specific targets of genetic diversity are not set other than for natural regeneration. The SFI system is even more vague, and does not set out criteria for maintaining or enhancing genetic diversity (AF&PA 2000). In contrast, the FSC is specific: "Diversity in the composition of plantations is preferred, so as to enhance the economic, ecological and social stability. Such diversity may include the size and spatial distribution of management units within the landscape, number and genetic composition of species, age classes & structure" (10.3, FSC 2000a).

At a national level, the FSC sets detailed targets for the minimum species diversity in any plantation. For example, line 365 of FSC UK guidelines (UKWAS 2000a) limits the proportions of different species within a wood. A maximum of 65% of the wood may be of a primary species, and only 75% of the wood may be composed of one species if it is the sole species suited to the site and the owners' objectives. The remainder of the wood must be a certain combination of native broadleaves, open space, secondary species or areas actively managed for biodiversity. However, this has resulted in oak (Quercus robur and Q. petrea) coppices in Wales, S.W. England and the Wyre Midlands having to be diversified despite being the result of decades of traditional management (Simon Pryor, pers. comm.). Similarly, in Germany, forests that are considered models of excellent SFM cannot gain FSC certification because beech (Fagus sylvatica) dominates the canopy, just as it would under natural conditions (Hans-Albrect Wiehler, pers. comm.). Critics of such an arbitrary figure are justified; no account of the natural forest composition is made. Moreover, multiple species composition is not always correlated with diverse flora and fauna – woods with a high proportion of English oak (Q. robur) will have much greater biodiversity than a dense plantation of mixed exotic conifers (Peterken 1992). Each situation should be examined on its own merits.

FSC UK guidelines, line 365, was recently clarified for a clonal poplar system. "The fact that a primary species comprises two or more different clones, varieties, provenances or origins does not alter the requirement for there to be no more than 65% (or 75%) of a species" (UKWAS 2000a). Yet it has been noted that these guidelines could be stretched, permitting a single poplar clone to comprise 65% of the forest (Simon Pryor, pers. comm.). Interestingly, this FSC UK standard is contrary to some other FSC national criteria. For example, FSC Sweden criteria 6.5.11 notes "cloned material is not to be used on a large scale pending an environmental impact assessment". It is likely this criteria has been established for FSC Sweden because Swedish (and German) government regulations prohibit the use of monoclonal plantations (Zobel 1993) unless they are arranged in a way that simulates a seedling population (Rod Griffin, pers. comm.).

It has also been noted that FSC certification could be awarded in plantations where diversification is merely planned, rather than after it has been achieved (Simon Pryor, pers. comm.). Indeed, the issue of 'planned improvement' has become very contentious and applies to other sectors of FSC certification: "...worryingly, many certificates appear to be awarded on the basis of hoped-for improvement in the management logging operations, rather than the actual good quality at the time of assessment" (Counsell 1999). If GM forests are to be certified they will have to meet these existing FSC criteria. It is clearly likely that GM would pose no additional threat to forest diversity beyond that already associated with plantation forestry.

ASEXUAL TRANSFER OF GENES

Antibiotic resistance genes have been used extensively in the generation of GMOs. Resistance genes are incorporated in target plant genomes, along with the desired gene sequence conferring a specific trait, as a means of identifying that transfer has taken place. Such sequences are often referred to as selectable markers. This has caused some concern amongst the scientific and 'environmental' communities; "Asexual transfer of genes from GMOs with antibiotic resistance to pathogenic micro-organisms, and/or suppression of mycorrhizae and other micro-organisms, arising from the use of GMOs with antibiotic resistance" (FSC 1999a).

The transfer of antibiotic resistance genes among prokaryotes is known to be common, although it may occur between bacteria and higher organisms (Istock 1991). For example, horizontal gene transfer is the most parsimonious explanation for the shared biochemical pathways of some microbes and plants (Strobel et al. 1994; Radmacher 1996). Moreover, gene homology has been found in Bradyrhizobium japonicum, and is probably the result of eukaryote to prokaryote gene transfer (Carlson & Chelm 1986). Host plant DNA has been found in the spores of the parasitic fungus Plasmodium brassicae in the laboratory, although whether this DNA is incorporated into the prokaryotic genome is not clear (Bryngelsson et al. 1988). The successful incorporation of DNA would require the recipient organism acquiring the transgenic trait, and then passing the gene to the population would depend of the fitness effects of the gene, subsequent selection pressures put on the gene, and the scale of gene flow to the recipient population. In all, there is a negligible likelihood of horizontal gene transfer but many potential mechanisms are proposed (James et al. 1998). If horizontal gene transfer were not so rare there would be extensive genetic similarities between plants and bacteria (Strauss 2000a). Indeed, spontaneous mutations giving rise to antibiotic resistant bacteria are several orders of magnitude more likely than asexual DNA transfer from GMOs (McHughen 2000, p.186). Although the potential effects of a possible transfer of antibiotic resistance may be very damaging, rational risk analysis on a case by case basis invariably finds it perfectly acceptable (Apel 2000).

Antibiotic resistance genes are not a prerequisite for producing GMOs; many GM-trees are being produced using alternative in vitro selection systems, particularly by Nippon Paper (Ebinuma et al. 1997). If the selection sequences are benign, GMOs would not warrant any fears about the transmission of antibiotic resistance, and hence should satisfy this aspect of certification demands.

Herbicide resistance genes

The FSC has three concerns over the introduction of herbicide resistance genes : increased weed resistance; increased herbicide usage; and transgene escape. FSC concerns are paralleled by some environmental NGOs; Section 2.3 of a WWF report raises concerns about "augmenting the selection pressure on target weeds ... facilitating herbicide resistance in the long term" (Owusu 1999). The concerns are valid, since herbicide resistance in weeds has already been documented numerous times in 'conventional' agriculture (Shultz et al. 1990).

The UK FSC standard notes "synthetic chemicals may be needed" (FSC 1999b). It also states that this section will be reviewed following the production of a DSS, developed under Section 5.2 of the analogous UKWAS certification system (FSC 1999b). The DSS is to aid in pest, weed and nutrient management decisions, and has yet to be published. However, a draft DSS document (UKWAS 2000b) is in circulation. It notes "guidance is based on the premise that

- Pesticide usage should not be the automatic method of first choice for controlling pests and weeds;
- There are problems for which research and experience have so far failed to find non-chemical remedies that do not entail excessive cost or pose a greater environmental risk than pesticides."

In other words, all options should be explored when dealing with a problem, and in some instances chemical methods are appropriate because they can be economically or ecologically advantageous. Continuing this argument, GM should be considered, for it may produce cheaper and more environmentally sound pest control (USDA 1999).

The DSS gives guidelines about assessing which methodology, whether chemical or otherwise, to apply to potential problems. It outlines parameters such as cost, effectiveness, toxicity and impacts of different practices. It continues "If the use of a chemical is necessary and a choice is available, evaluate the risks of environmental damage". Further, "...reference should be made to 'Environmental Impacts' which gives general information and a table to assist in choosing the material with the least impact". Essentially, the DSS gives a permitted list of options, chemical or otherwise, and asks the forest manager to make a value judgement for best practice and damage limitation: "Users should decide ... using their professional judgement" and "determine the range of suitable pesticides, dose rates and application patterns from the tables and supplementary literature ... make an assessment on the likely effects of the pesticide using the table in this section". Thus, each individual situation needs consideration in the field, from which an appropriate course of action can be taken.

One could argue that in many instances GM fulfils DSS requirements, for the "aim should always be the minimum quantity of herbicide required to give the desired degree of control". An ERS/NASS Agricultural Resource Management study (USDA 1999) on herbicide treatments on GM herbicide resistant cotton and soybean compared application rates to all other seed technologies, and found that 50% to 60% of spayed crops had significantly lower herbicide application at the 5% level (USDA 1999). Further USDA research has shown that between 1997 and 1998, U.S. national pesticide treatments were reduced by 6.2% and was accounted for by the introduction of Bt resistant crops (Fernandez-Cornejo et al. 2000). Reduced herbicide applications may also lessen the selective pressures on weeds.

The DSS report continues "herbicides that have low toxicity should be favoured" and "in many cases a carefully directed spray of broad spectrum product will be the most effective option and offer the least risk to non-target species". These are often cited as pro-GM arguments. The introduction of a herbicide resistant transgene such as Roundup Ready® confers resistance to glyphosate, and has been considered as one of GMs greatest successes (Tzfira et al. 1998). The DSS states glyphosate (Roundup Pro Biactive®) is "not hazardous, not toxic to invertebrates, not harmful to aquatic life, and has low selectivity". It is a safe, broad-spectrum herbicide, considered environmentally benign because it breaks down quickly into non-toxic end products. If used properly, a Roundup Ready® gene could lessen the environmental impact of weed control; over the top spraying becomes feasible (Dickson & Walker 1997), enabling post-emergence treatment when weeds are vulnerable (Rogers & Parkes 1995), and a reduction in herbicide usage because the "use of pre-emptive weed control is often more effective than dealing with greater problems later" (UKWAS 2000b). A recent study confirmed these benefits; Roundup Ready® agricultural crops enabled the substitution of most synthetic herbicides with glyphosate. "The herbicides that glyphosate replaces are 3.4 to 16.8 times more toxic .... and persist in the environment nearly twice as long" (Fernandez-Cornejo et al. 2000).

Indeed, conventional weed treatments, such as tilling and mowing, can be very damaging to the environment. The DSS report notes that "Mowing is largely ineffective as a weed control measure" and "often needs to be repeated several times a year for many years". Furthermore, "mowing creates a grassy weed flora that is harmful to trees" and "can result in soil compaction ... and pollution from exhausts and spillage of fuel and lubricants". The reduced need for tillage control of weeds can lessen erosion (James et al. 1998).

The FSC recognises that, under some conditions, herbicide use is warranted. Chemical use is a judgement made by the forest manager. If he or she justifies usage soundly, having considered all possible alternatives, then certification follows. This rationale could be extended to GM. There are many possible applications of transgenics. The specific trait and its proposed use should be examined, just as for other technologies. Furthermore, substantive equivalence should be cited; the DSS goes on to state "there are some situations in which complete removal of all other vegetation over the whole site may be considered". Despite the fact that such a radical measure could have a significant effect on a site, it has been judged to be acceptable. It is highly unlikely that GM would result in removal of all other vegetation from a site, yet it has not been judged to be acceptable.

Insect resistant GMOs

The FSC consider a potential hazard of GM trees to include "increased resistance of target insect pests, and/or deleterious effects on natural enemies of the target insects, and/or deleterious effects on insects such as butterflies, pollinators, soil microbes, arising from the use of GMOs with insect resistance" (FSC 1999a). These concerns are thought to stem largely from U.S. studies on the effects of GM-corn modified with Bt transgenes. Two studies have reported negative impacts of Bt toxins on the monarch butterfly, Danaus plexippus; Losley (1999) and Hanson & Obrycki (2000). These papers have since been criticised as unscientific or taken out of context by the media (Shears & Sheldon 2000), but have generated public apprehensions. Other concerns are voiced by NGOs; Section 2.4 of a WWF report considers insect pesticide resistance, noting "Bt crops will augment the selection pressure placed on target pests and this will inevitably lead to an increased frequency on Bt-resistance genes within the insect's gene pool" (Owusu 1999).

Such misgivings over continual plant insecticide production are valid. Bt sprays have been used for decades in conventional and organic agriculture, and are considered "a safe and effective bio-pesticide" (Greenpeace 2000), despite scientific research questioning the safety of Bt spays (Swadener 1994). For forestry operations in the US and Canada, Bt is often the only insecticide that is sanctioned. It has also been widely used in Europe; in Spain Bt is frequently applied against gypsy moth Lymantria dispar (Speight & Wainhouse 1989). However, the continued exposure of pests to Bt toxins has selected for many resistant insects (Tabashnik et al. 1990; Talekar & Shelton

1993; Tabashnik 1994; Bauer 1997; Tang et al. 1997; Speight et al. 1999). Such resistance is thought to have developed through continued exposure to sprays on non-transgenic crops, and where these sprays have persisted in the soil following application (Saxena et al. 1999).

Despite ten years of proven insect resistance in the field, Bt sprays remain on the market and are an acceptable practice under rigorous 'organic' guidelines (EU 1991). This may lie in the perception that Bt is a 'natural' product, produced by bacteria. In fact, Bt is not an obligate pathogen. Vegetative cells can be induced to sporulate and produce a crystal containing toxic proteins (? endo-toxins) when they are grown in sub-optimal regimes in fermentation chambers (Speight & Wainhouse 1989). The FSC has misgivings over the use of transgenically introduced Bt toxin, despite studies showing that Bt transgenic plants can often provide a pesticide delivery system that manages insect resistance better than pesticide delivered through Bt sprays (Roush, 1994). Whilst it is undecided if pest control using spray Bt as a whole bacterium is dangerous (Swadener 1994), it may be safer to use Bt transgenes incorporated by GM, and so reduce the probability of gene transfer between bacterium (Helgason et al. 2000).

Concern about the "build up of resistance to natural biological controls or the chemicals used to control them" (Soil Association 1998) is legitimate and of considerable importance. Pests acquiring resistance from elevated exposure to selective controls could be problematic for future generations. Insect resistance to Bt will eventually develop, GM or no GM, because Bt sprays have been used extensively by organic farmers since the 1950's (McHughen 2000). The same is true of chemical pesticides, used by conventional farmers for decades. In fact, given that the precision of delivery is increased with GM, there may be a lower likelihood of the development of insect resistance to Bt toxin with GM crops, especially if refugia of non-GM trees are used (Roush 1994).

The use of many chemicals is permitted under UKWAS, and thus FSC, guidelines (UKWAS 2000b). Some have potentially severe impacts, and it is these chemicals with which GM must be compared against. For example, permethrin is used for the control of Hylobius abietis, a weevil that can devastate restocked stands (UKWAS 2000b). Described as "low selectivity, extremely dangerous to aquatic life, dangerous to invertebrates and toxic to bees (do not apply when flowering vegetation present)." It "may damage non-target vegetation", and has been "identified as a potential endocrine disrupting chemical by IEH (UK Institute for Environmental Health) and EA (Environment Agency)" (UKWAS 2000b). Yet under UKWAS and UK FSC standards, use of permethrin is acceptable. This practice is contrary to FSC International standards (FSC 2000a), a position expounded by the Swedish FSC standard: "substances, including chemical pesticides and herbicides in the Chemicals Inspectorate's class 1 & 2 that are harmful to the environment & health shall not be used for the treatment of forest land. Permethrin treatment is currently exempted (until and including 1999)" (FSC 1997). Why would an exception be made for a chemical that otherwise contravenes FSC standards. It is the only means of controlling a potentially devastating pest. Other chemicals e.g. granular carbosulfan, and remedial options, e.g. weevil guards, are feasible but are more expensive . Moreover, "plants may require additional treatment(s) after planting, especially if small plant sizes are used" (UKWAS 2000b). Permethrin is certainly the cheapest, most commonly utilised and widely vocated practice. It may deemed acceptable because it has been used for years.

GM could conceivably provide a more 'ecologically friendly' alternative. Bt toxins are very specific insecticides, and target only susceptible species (Raffa et al. 1997; Speight & Wylie 2001). A Bt toxin specific to H. abietis could be inserted into the plant, just as has been done in Populus against the beetle Chrysomela tremulae (Moffat 1996). This would target only the pest, and possibly closely related species, which both predated the plant; a very precise application of insecticide. In contrast with the current practice of using permethrin, such an application of GM may be justified under DSS guidelines: "the use of selective pesticides may offer less impact than some non-chemical methods that are not species specific" (UKWAS 2000b).

Although not taking a specifically GM stance, 5.2.b of the PEFC standards states "inappropriate use of chemicals or other harmful substances or inappropriate silvicultural practices influencing water quality in any way should be avoided" (PEFC 1998). 6.3.3 of Annex 2 of PEFC Sweden is more succinct: "the use of chemical treatment is minimised and only used when other suitable methods do not exist" (PEFC 1999). In light of the earlier DSS discussion (pages 13-15), which describes appropriate methods of soil, weed and pest management, it is fair to comment that using chemicals can be justified in certified forests. One could thus argue for GMO certification following a rigorous assessment of impacts and risk. Interestingly, PEFC Sweden goes on to state "It is especially noted that permethrin is allowed during an exception period".

Whilst potentially reducing environmental impacts, transgenic plants producing pesticides do not allay the fears over increased pest resistance. The introduction of single genes may be the easiest approach to resistance, but single genes are susceptible to shifts in the gene pool following selective pressures, and are thus vulnerable to mutations within pest populations (Pimentel et al. 1989). This is particularly pertinent for tree crops, which have years of exposure in a single rotation. In agricultural crops monogenic resistance is overcome in a few years (Wilcox et al. 1996), so pests could devastate tree plantations in mid-rotation. This has been recognised by most proponents of GM, who recommend a very selective use of transgenic pest resistant trees until more genes of diverse function have been identified (Strauss 2000a). However, every gene system and not all simple resistance systems will be overwhelmed rapidly. For example, linkage analysis in loblolly pine (Pinus taeda) showed that resistance to the rust Cronartium quercuum was monogenic and has been successful despite more than 40 years exploitation as a commercial crop (Wilcox et al. 1996).

If exposure to a pesticide could be reduced, the selective pressure on target species might decrease, hence reducing the rate at which a species attains resistance. This could be achieved by linking the expression of inserted genes to inducible promoters. A level of predation could then be tolerated before the trees transgenic defences were activated (Strauss 1998). Furthermore, since many pests attack trees at a specific age (usually juvenile), it could be possible to link transgene expression to developmental stage. This would also reduce the exposure of pests to a toxin. The most prudent compromise seems to be the use of transgenic resistance as a supplement to existing quantitative resistance developed through conventional breeding (Burdon 1999). Thus, coupled with Quantitative Trait Loci (QTL) analysis using Marker Assisted Selection (MAS), GM could actually help ensure more lasting pest resistance (Burdon 1999).

FSC Criterion 6.6 notes "management systems shall promote the development and adoption of environmentally friendly non-chemical methods of pest management and strive to avoid the use of chemical pesticides". As noted in the DSS, judgements about pest management must be made by the forest manager and tailored to site specifics (UKWAS 2000b). Importantly, GM could be viewed as an effective non-chemical method. A major motivation for transgenic research is that GMOs are expected to have fewer ecological impacts than use of current chemical practices. English Nature recognise that the use of crops modified for insect tolerance may have potential benefits for farmland wildlife, particularly if their use results in "better targeted or lower impact of agrochemicals" (English Nature 2000). Thus a GM tree producing allopathic or insecticidal chemicals could fulfil criterion 6.6.

Lignin modification

Trees modified for lignin, and indeed other traits such as pest and herbicide resistance, will only be of advantage using short rotation crops such as poplar, eucalypts and some pines (Griffin 1996; Strauss 2000a). Indeed, Jonas Jacobsson of AssiDomän stated "There may be an economic advantage [of using GM] if you have a plantation with a short rotation" (3C Associates 2000).

The modification of lignin biosynthetic pathways could improve the suitability of trees for pulping (Pullman et al. 1998). This can be achieved by reducing lignin content or altering the biochemical composition of lignin; both mechanisms may reduce the need for chemical inputs to treat the pulp. For example, in transgenic tobacco the modification of lignin biosynthetic pathways reduced the Kappa value by 15% (Halpin et al. 1994), i.e. 15% less chemical bleaching was required to whiten the pulp. Enhanced lignin extractability has been reported with GM poplars (Baucher et al. 1998; Whetton et al. 1998). The potential benefits are substantial, with estimates suggesting that pine cad mutants could save $4 per tonne of paper produced by the Kraft process (Jeremy Brawner, pers. comm.). Indeed, a 1% reduction in timber lignin content would increase pulp yield by 1% - 1.5%, and could save the global paper industry billions of dollars (3C Associates 2000). Moreover, reduced inputs of energy, chlorine and other hazardous chemicals could lessen environmental impacts proportionately.

Booker and Sell (1998) suggest that lignin modification could compromise the structural integrity of trees and reduce their natural defences, whilst the FSC is concerned about "adaptation and pest resistance of trees, rate of decay of dead wood, and soil structure, biology or fertility, arising from use of GMOs with modified lignin chemistry" (FSC 1999a). However, it appears possible to modify the lignin content of trees without compromising plant viability (MacKay et al. 1997; Franke et al. 2000), and it is likely altered lignin composition will have very minor consequences (Dickson & Walker 1997).

Clone 7-56 of loblolly pine (Pinus taeda) is a naturally occurring mutant with a modified lignin biosynthetic pathway (MacKay et al. 1997; Ralph et al. 1997). Conventional tree breeders long ago identified 7-56's value as a parent in the production of trees with improved growth and pulping characteristics and thus incorporated it into their breeding programmes. Consequently, 7-56 progeny are planted on a massive scale throughout the southern states of the USA (Jeremy Brawner, pers. comm.). There have been no reported negative side effects of using this clone or its progeny, and no certification programmes have an issue with 7-56, having certified several thousand hectares (Jeremy Brawner, pers. comm.). MacKay et al. (1995) showed that mutation of the cinnamyl alcohol dehydrogenase gene (cad1-n) is responsible for 7-56's modified lignin. Plants that have an equivalent phenotype to the cad1-n mutant have been generated for tobacco and poplar (Halpin et al. 1994; Lapierre et al. 1999). Thus the same phenotypic effect can be derived by both conventional and GM methods. It seems paradoxical that a conventionally bred CAD null tree can be accepted and yet CAD-deficient mutants generated through GM cannot. They have the same phenotype and are thus very likely to have the same ecological impacts.

This example illustrates a problem for certification with respect to the use of GM to alter lignin. That is, 7-56 homozygotes and CAD-deficient plants derived through GM are phenotypically equivalent, discriminated simply by method of production. Moreover, the FSC's concerns over lignin modification altering "rate of decay of dead wood, soil structure, biology and fertility" conflict with other FSC-accepted practices. Conifers, such as pines, and angiosperms, such as eucalypts, can be interchangeably planted on the same land. Their very different lignin and wood chemistry, as well as other chemical differences between these diverse taxa, may bring about ecological changes greater than anything possible via modification of lignin through GM per se. Moreover, the wood chemistry of exotic species is thought to effect the ecology of plantations far more than lignin modification of a native species. It is this that makes some believe the FSC has "no place bringing lignin modification as a rational concern" (Strauss 2000b) regarding the certification of GM trees.

Transgene escape

Widespread anxiety has been expressed over the possible contamination of wild populations by transgenic plants, which could compromise natural fitness or introduce some components of fitness that would increase a species weediness potential (Rogers & Parkes 1995; van Raamsdonk & Schouten 1997; Burdon 1999; Ellstrand et al. 1999). The FSC have detailed their concerns over transgene escape , and make particular reference to the potentially negative impacts arising from the spread of herbicide resistance .

Assessing transgene escape is very complex (Kjellsson 1996; Ammann 1999). In brief, considerations must be made regarding: (1) the likelihood of gene flow; (2) the probability of gene establishment (introgression) (Ellstrand et al. 1999); and (3) the impact of an introgressed transgene, which will depend on its effect on the ecological factors regulating the recipient plant population (Crawley et al. 1993; Dale 1994; Schmitt & Linder 1994; Gray & Raybould 1998). Assessing and quantifying these variables draws on theoretical ecology, comparisons with introduced exotic species and recently conducted experimental work. Trees present additional obstacles for analysis because of their long generation times, large size and potential for long distance dispersal of pollen and seed (DiFazio et al. 1998). Thus each transgene and its application must be examined in isolation. These factors have yet to be assessed, since not a single GM tree has yet received approval to reach sexual maturity in a field environment (Griffin 1996). Consideration is given to gene flow, and increased 'invasiveness'.

Gene flow

To date, most studies have focused on gene flow via pollen in agricultural crops (van Raamsdonk & Schouten 1997). This follows a rational call to assess each species independently due to differences in pollen biology: wind-pollinated plants will disperse their pollen over much greater distances than insect-pollinated species. Pollen may flow into the gene pool of a nearby species, by either outbreeding depression (fitness reduction following hybridisation) or genetic assimilation (dilution of the genetic integrity of the wild species until it is effectively assimilated into the crop species, Ellstrand 1992). This is not restricted to GM plants, but is a potential problem for any plantation, particularly if it constitutes a narrow gene pool since similar genes could 'flood' neighbouring populations. For example, hybridisation with non-transgenic crops has been implicated in the extinction of at least five wild relatives of food crops (Small 1984) and the Californian walnut (Juglans hindsii) (Ellstrand 1992) .

The more taxonomically distant an exotic tree is from native species, the lower the chance of gene transfer. Conceivably, if sufficiently distant exotic species of GM trees were used, there would be no risk of transgene passage to native plants. They could thus be certified. Conversely, the more closely related a crop tree is to indigenous wild relatives, the greater the likelihood of gene flow. Thus, in the UK, crops have been classified based on their potential to hybridise with wild species (Rogers & Parkes 1995); black poplar (Populus nigra) and Scotts pine (Pinus sylvestris) have been placed in the highest risk category. Physical proximity is also important; pollen dispersal experiments have confirmed a negative exponential or leptokurtic range of pollen distribution (Scheffer et al. 1993). However, gene flow in poplars has been demonstated over distances of at least 10 Km (Strauss et al. 2000c). This research exemplifies the potential for transgene spread; limited gene flow is inevitable if GM plants are grown close their relatives (van Raamsdonk & Schouten 1997; Ellstrand et al. 1999; McHughen 2000, p.166). It is also noted that "it is a clear mistake to place emphasis on pollen-based gene flow and nothing on the more obvious [sic. seed] route for gene escape" (McHughen 2000, p.166), for this enables gene flow through time as well as space. It is probable that legislation will require sterility in GM trees (Burdon 1999).

If genes do manage to pass between related species, several modifications are predicted to give plants a selective advantage, including enhanced tolerance to environmental factors, such as salinity, and pest resistance (Rogers & Parkes 1995). Herbicide resistance is only likely to confer a selective advantage to plants if they are exposed to herbicides. However, plants carrying these traits may be disadvantaged due to the metabolic costs of synthesising proteins in the absence of a selective advantage. Experiments with perennials generated mixed results, suggesting there are no costs associated with resistance genes (Lavigne et al. 1995) but weediness is not increased (Crawley et al. 1993; Snow et al. 1999). However, research is yet to be performed on trees, and is also needed to determine if there is a greater risk attributable to GM transgene spread than to conventional gene spread. However, one of the problems facing regulators and users of GM plants is the quantification of small risks from such experiments, and then scaling them up to plantation scale (Rogers & Parkes 1995). Large-scale releases of pollen will result in a larger flow of genes to wild relatives. Such genes may increase in frequency in the wild population irrespective of their fitness effects, such that a large scale plantations may be sufficient to overcome any deleterious fitness effects of the transgenes (Dale 1994; Gliddon 1994; van Raamsdonk & Schouten 1997). This issue is also not unique to GM plantations, but to any large scale plantation.

The risk assessment of gene escape also depends on the nature of the gene. Again, this is equally true of conventional and GM plants. For example, in agriculture 'Smart Canola' has been conventionally bred to express resistance to two herbicides (Concar 1999). Such plants present very similar risks as GM ones, and should therefore be subject to the same risk assessments, for "if the concern over gene spread is valid, the method of initial production is irrelevant" (McHughen 2000, p.113). Thus, the certification requirements imposed on any plantations should be examined.

Most certification systems make specific reference to utilising appropriate provenances, varieties and species . These criteria, particularly those of the PEFC, are designed to address principally the concerns over gene spread between endemic species. Some criteria are very specific, for example 6.3 209 FSC UK requires that "species diversity is maintained and dilution of the local gene pool is minimised" (FSC 1999b). Indeed, FSC UK set stringent requirements on total species composition. However, the caveat 'appropriate' and similar dispensations appear in most of these criteria: "Seed of local provenance is used wherever it is available and considered appropriate for planting and restocking of native species" (9.2 335 FSC UK); "for reforestation and afforestation, origins of native species and local provenances that are well adapted to site conditions should be preferred, where appropriate." (4.2b, PEFC 1998). These exemptions recognise that sustainable forestry is a balance between ecological and financial objectives, and ultimately dependent on numerous variables for which it would be impossible to specifically legislate.

Certification may be awarded for planned improvements, and forest managers may make extremely literal interpretations of the criteria. Whilst the criteria aim to prevent subjectivity, ultimately the interpretations of individual assessors will have a role, which may explain why some certification standards have been criticised (Counsell 1999).

Increased 'invasiveness'

"Considering the analogy with exotics has been a helpful one in regulating the release of GM" (Nuffield Council on Bioethics 1999); this is because the community level effects of transgenes acquisition by wild relatives have been hypothesised to be similar to those of introduced species (Schmitt & Linder 1994; Williamson 1994; Rogers & Parkes 1995; Rissler & Mellon 1996). Most seriously, the competitive superiority of GM plants, as a consequence of an acquired transgene, may lead to the exclusion and extinction of the native plants (van Raamsdonk & Schouten 1997); "it is the traits that increase competitive behaviour that are of primary concern" (Nuffield Council on Bioethics 1999). The effects of increased competitiveness can cascade through the ecosystem (Rissler & Mellon 1996), for example, in the United States, 42% of the species on the threatened or endangered species list are at risk primarily because of non-indigenous species (USDA 1999) costing the US economy an estimated $138 billion a year (Pimentel et al. 1999). However, some prominent GM scientists "do not believe that transgenics have properties even remotely similar to invasive exotics, and should not be considered along with them for scientific or regulatory purposes". Exotics form the mainstay of industrial plantations in many countries, and total 98% of the US food system (Pimentel et al. 1999). This is because they grow so well; 10.4, 363 FSC UK notes "...native species are preferred. Exotic species are only used where they will substantially out perform native species in terns of meeting the objectives of plantations" (FSC 1999b).

In New Zealand 6% of the country is covered in plantation forest. This generates the majority of US$1.42 billion of timber product exports (FAO 2000b), which accounts for 5.5% of national export earnings (ODCI 1999) . Most of the plantations are of the exotic Pinus radiata, which has been intensively selected, converting the poor wild form into a straight-trunked, fast-growing tree crop (Chilvers & Burdon 1983). Pinus radiata can be highly invasive (Cronk & Fuller 1995), and is considered to be a 'significant problem' in scrubland, forest margins, sand dunes, open land and short and tall tussockland (Williams & Timmins 1990). Fletcher Challenge Forests recently decided to opt for FSC and ISO certification of their New Zealand forests, in order to maintain their position in U.S. markets (Kelly 2000). Earlier FSC certifications in New Zealand, such as the Rayonier's 34,000 ha Southlands Estate, have not considered exotic species inherently problematic. As FSC 6.9 states, "The use of exotic species shall be carefully controlled and actively monitored to avoid adverse ecological impacts". The situation is similar in Scotland. 6.1 173 FSC UK states "operations planned with consideration of: the spread of invasive species across the forest boundary in either direction". Thus, it could be argued that the certification of exotic and/or invasive species is possible when contingencies exist for potential negative impacts.

The issue of invasive exotics is also pertinent in South Africa, where timber and pulp production are almost entirely dependent on exotic species – pines, eucalypts and Australian acacias ('wattle') (von Maltitz 2000). Many of these species have been demonstrated to be highly invasive, particularly in the Fynbos (Cronk & Fuller 1995) and veld-grasslands (Cooper 1999). Indeed, SAPPI have a special eradication programme for alien invasives. Yet three forestry companies have hundreds of thousands of hectares of FSC certified exotic, potentially invasive, forest (FSC 2000c; Von Maltitz 2000). This has drawn criticism from environmental groups in South Africa such as WESSA, who have expressed concern about the credibility of the FSC over their policy surrounding exotics (Murphy 1999).

If the potential impact of GM trees is to be analysed, the risks must be considered against current policies for exotic species. The introduction of non-native species is specifically warranted by the FSC. For example, the great spruce bark beetle Dendroctonus micans is a pest of exotic spruce plantations (Speight & Wainhouse 1989) and costly crop losses can be controlled by the exotic predatory beetle Rhizophagus grandis. Thus, 6.9 270 FSC UK states the "use of non-native biological controls such as Rhizophagus grandis may be desirable to control non-native pests". The use of this exotic beetle reduces economic costs and insecticide application. Its interactions with the environment have, and never will be, fully elucidated, but the benefits are deemed greater than the potential impacts. These arguments ring true for GM.

RESTRICTED ACCESS TO ADVANTAGES

The development costs of GM trees is great. Thus, even large forestry and biotechnology companies are collaborating in projects such as ArborGen . However, investing millions of dollars in this field is risky, and only deemed acceptable since patent protection provides a temporary monopoly over new products, enabling companies with the leading technology to dominate their sector. This has raised concern amongst NGOs, who fear poorer growers will be unable to afford increased capital expenditure, and thus have restricted access to the advantages of GM. The FSC acknowledge this anxiety . However, justifications for sharing of intellectual property will be very difficult, and out-of-line with the increasing harmonisation of global patent protection.

However, those pursuing GM technology argue that the higher cost of GM-trees is a demonstration that the trees have additional value; growers will only purchase new products if the additional value is worth the higher price. For example, a traditionally bred hybrid maize seed is purchased by most U.S. farmers since it generates more income, justifying the extra initial expenditure (McHughen 2000, p.192). This is despite having to buy new seed each season. In contrast, the GM tomato, Flavr-SavrTM, failed because it was too expensive and the flavour did not warrant consumers spending the extra money (McHughen 2000, p.257-258). However, many small growers, especially in developing countries, cannot afford capital outlay. As illustrated by the Green revolution, additional costs favour larger growers. However, this issue is not specific to GM, and it would be wrong to consider GM in isolation.

If we consider current practice, many growers already purchase their seed or seedlings from specialist growers; in vitro generated clones and advanced breeding programmes are beyond the scale of all but large forestry companies or specialist nurseries. If foresters wish to plant improved varieties, invariably they buy from these suppliers, since their access to improved varieties is already restricted, regardless of whether the product is GM. Although certification schemes make specific references to using local provenances where appropriate (PEFC 1998 4.2b; FSC 2000a 9.2), the onus is on genetic diversity. Reference is not made to restrictive or monopolised access to quality seed.

Certification is espoused as a means of ensuring socially responsible and sustainable forestry. Yet critics have noted that becoming certified imposes additional costs upon growers (Centero 1998). This additional expenditure, particularly that of the FSC system, is seen by some to exclude small growers (Counsell 1996), especially in developing countries (Banahene 2000). In a market increasingly concerned with a reliable supply of a standardised product, certification makes it even more difficult for small growers to gain access to markets (Counsell 1996). Some have considered a major driving force behind certification to be large companies intent on expanding their market share (Counsell 1996). This may explain why 66% of forests certified under FSC guidelines have been by large industrial enterprises (Thornber 1999). Although thc certifiers have acknowledged the difficulties faced by small growers, and have tried to develop group systems, the additional expense and limited supply of certifications is restrictive.

Reduced biodiversity from sterile trees

Due to the potentially negative effects of gene flow, sterility may be required for the release of GM trees (English Nature 2000). The ecological impacts of large sterile plantations has raised concerns (Dr George McGavin quoted in Tickell & Clover 2000) over reduced biodiversity. The FSC is specifically concerned about the "Reduced biodiversity of organisms dependent on flowers and fruits, arising from use of sterile GMOs" (FSC 1999a). Moreover, the FSC has doubts over the ability of trees engineered for sterility to prevent gene flow to native plants .

Engineered sterility is seen mainly as a risk reduction measure to minimise negative effects of transgene flow and genetic erosion. However, sterility can offer other benefits. Reproductive growth requires between 15% and 30% of a trees energy (Owusu 1999). Thus, suppression of reproductive tissues could channel more resources into vegetative growth and thus could greatly increase productivity. Furthermore, it could reduce airborne allergens.

Sterility can be conferred in two ways. The first involves the suppression of the floral genes that are essential to produce fertile gametes. This can be achieved through either antisense suppression of gene expression or homology-dependent gene silencing . The principal disadvantage of these approaches is that native genes or highly homologous equivalents are needed. This greatly increases the research time and expense of modification. Furthermore, floral development is very complex, and thus redundancy would have to be built into the system in order to ensure stability throughout a plant's life, and would require many genes to be used. For example, LEAFY (LFY) in Arabidopsis controls a developmental switch to convert lateral shoots into flowers (Weigel & Nilsson 1995). The main stem must acquire competence to respond to LFY, and this capability increases during the life cycle (Weigel & Nilsson 1995). Thus, genes prior to LFY control LFY expression, and include promotors of flowering such as COSTANS and inhibitors of flowering such as TERMINAL FLOWER. Research using PTFL, a poplar LFY homologue, indicates that these promotors and inhibitors are in turn governed by a complex system of genes that respond to day-length (Rottmann et al. 2000). Moreover, genes downstream of LFY control precise floral organ development

(Coen & Meyerowitz 1991). By targeting different genes in this complex system redundancy may be ensured.

Suppression of floral genes is attractive for modification since fertility is impaired at various stages of floral differentiation (Melian & Strauss 1997), which enables selective sterility to be generated in tissue-specific regions. For example, modifying structural or catalytic proteins essential for pollen formation could generate male sterility. Whilst safe-guarding against gene flow via pollen, breeders could still cross female flowers with pollen producing trees. However, such modifications of pollen producing genes may not enhance vegetative growth, because expression is late in floral development. Moreover, the problems of seed production and gene flow via seed still exist, and the FSC is specifically concerned about partial sterility that does not entirely prevent gene flow (FSC 1999a).

The second means of generating sterile trees would be through the expression of cytotoxin genes in a floral tissue-specific fashion. This would disrupt or ablate organ-specific tissues and since early and frequent floral-specific expression, floral homeotic gene promotors are probably best suited for engineering complete sterility, whilst enhancing growth (Meilan & Strauss 1997). However, occurrences of instability in transgene expression have been reported (Denis et al. 1993), and it is these instances that concern ecologists and the FSC about transgene spread. Trees remain in situ for many years, and are seldom closely monitored. A breakdown of sterility could go unnoticed for many years and allow GM genes to escape to wild populations.

If sterility is an important aspect of gene containment for regulatory and certificatory purposes, it is critical to demonstrate that sterility is maintained under a range of conditions and development stages. This requires non-contained field trials. It will take many more years of work to develop guaranteed sterility systems (Strauss 2000a). Until then, existing sterility systems will be a risk reduction measure, and require an acceptance of small transgene releases. However, the problems of time-scale, massive pollen releases and the rarity of the breakdown in sterility makes monitoring difficult and costly. Coping with a breakdown in sterility and the potential consequences of transgene releases could be impossible. However, equivalence must be invoked when considering the ecological aspects of sterile plantations. If the technology is regulated properly and used with consideration, there is no reason to expect a negative consequence of failed sterility transgenes. Furthermore, it is unlikely the consequences of failed sterility will be as dramatic as those described for Pinus radiata introduction (Cronk & Fuller 1995).

Currently, large plantations of exotics are certified. The productive areas of these forests are often highly effect environments, with a biological composition radically different to the wild systems they replace (Peterken 1992). Certification systems acknowledge the detrimental impacts of exotic species, but agree the increased yield from non-natives justifies their use. Thus they aim to ensure ecosystem management to maximise beneficial activities and minimise adverse impacts. Equally, these ecosystem management practices could be applied to sterile trees, since the potential for increased wood production is great and certification systems could justify greater proportions of the forest dedicated to ecologically beneficial practices. GM is likely to be used with exotic species in short-rotation, commercial plantations (Griffin 1996; 3C Associates 2000; Strauss 2000a). The ablation of reproductive organs, and consequently the loss of reproductive food matter to organisms in these plantations, would represent only a small impact when contrasted to those already imposed through the use of exotics. The very reason exotics grow so prolifically is thought to be their unpalatablity to a significant number of pests (Speight 1989).

General concerns

The FSCs concern over "Reduced adaptability to environmental stresses, changes to interaction with other organisms, and increased weediness or invasiveness, in GMO trees with new features" (FSC 1999a) encapsulates all their previous points. Reference to the preceding sections should be made to discuss these specific issues. However, treating such issues as disparate overlooks the potential problem that these factors may compound and interact. The resultant gestalt could be unpredictable and pose potential problems.

EVALUATING THE RISKS

PREDICTABILITY AND INSTABILITY

Although not expressly voiced by certification systems, the unpredictability of genetic transformation has raised concerns in the scientific community. This is because the "Insertion of a novel gene can have a collateral impact on the rest of a hosts genome, resulting in unintended side effects" (Owusu 1999). New genes that could conceivably disrupt the metabolism of an organism will usually produce transgenics that are lethal. However, experience with maize showed that a cytoplasmic male-sterility factor led to the breakdown of resistance to the rust fungus Bipolaris maydis (Levings 1990). Although the technology used in this example is dated, genetic instability could produce many unintended effects that only manifest themselves years after deployment. The vast majority of transgenes are stable but there have been instances of instability, including altered patterns of gene expression (van der Hoeven et al. 1992) and the failure of engineered sterility (Denis et al. 1993).

Instability may be induced by high temperatures (Broer et al. 1992; Meyer et al. 1992; Walter et al. 1992), which are thought to alter methylation (Meyer et al. 1992). This illustrates an important principle. Since many genes are environmentally and developmentally regulated, the physiological state of plants can have an important effect on transgene expression, thus, the risks of collateral alterations to stress-activated genes "cannot be anticipated until the stress response is actually triggered" (Owusu 1999). This is of special import for forest trees because they are exposed to environmental fluctuations for much longer periods than agricultural crops. Moreover, a time lag could make the problems of instability more acute because the widespread planting of a particular cultivar could be spaced over many years. A greater time frame also has a bearing on the uncertain risk of novel gene mutation.

However, some have pointed out that in 1999 alone GM crops covered 45 million hectares (100 million acres) in North America, and there are no documented dramatic or adverse or unexpected effects from any of these plants (McHughen 2000, p.190). Studies with poplars have shown stable gene expression for several years (Strauss, pers. comm., 2000). The rigorous testing and regulatory policies prior to commercial GMO release are considered effectively to remove unstable cultivars. Extensive quantities of data are gathered in trails, and the quantified risks satisfy assessors and government legislators, although they may not allay the concerns of some groups who question the independence of assessors. Simple steps could also be taken to lessen the potential pitfalls of instability. Just as for risk aversion in clonal plantations, a variety of transformants could be planted. These could have a number of transgenic insertions with multiple copies of the desired transgenes (Burdon 1999). Again the debate centres on risk analysis, and as such should focus on the specific GM trait.

It is important to note that "all living things are subject to natural genetic instability" (McHughen 2000, p.189). Pieces of DNA called transposable elements can move directly from one chromosome site to another (Fedoroff 1992), multiply, and occasionally re-arrange neighbouring DNA sequences (Alberts et al. 1989). They are thought to make up 10% of higher eukaryote genomes, and can affect gene regulation (Alberts et al. 1989). Such inherent instability is in present in all plants. Genetic changes induced by transposable elements are accommodated by phenotypic plasticity. As such, GM would not present any new phenomenon of instability.

Risk assessment

Some FSC supporters are against the deployment of GMOs (Greenpeace 2000), whilst other supporters call for a moratorium (Owusu 1999). The spectrum of potential GM applications has lead many bodies to recommend that "regulators should explore the pros and cons of adopting a more explicit risk / benefit assessment" (Nuffield Council on Bioethics, 1999). Risk assessment can be controversial, reflecting the important role that both science and judgement play in drawing conclusions about the likelihood of effects on human well-being and the environment; contention often arises from incomplete knowledge. To make an effective risk management decision, stakeholders need to know what potential harm a situation poses and how great is the likelihood that this harm will be realised. Each application should be examined in isolation; however, the hazards of GM cannot be properly evaluated (Apel 2000). While the extent of exposure can be estimated, there are no responses to exposure. Contemporary risk models must thus make assumptions (Burdon 1999), and herein lie the seeds of dispute. Those against GM extol the precautionary principle based on our lack of knowledge, whilst others maintain negative assertions can never be proven, and GMO deployment should be based on comparisons with current practices. This has lead some to argue that the "anti-biotechnology contingent has made the lack of scientific knowledge part of a strongly subjective standard of risk assessment, and coupled it with a confirmation bias " (Apel 2000). Countering this, some of the scientific community believe "... the simple principle of genetic modification spells ecological disaster. There are no ways of quantifying the risks... ...The solution is simply to ban the use of genetic modification." (Narang 2000). This dichotomy of views may rest with philosophical beliefs surrounding 'natural' procedures and the role of multi-nationals in dictating much of biotechnology research in forestry. As such societal interests play a strong role in moulding the perceptions of GM. Reconciling these different perspectives may be very difficult.

Proving GM is non-hazardous requires a negative proof, and science cannot corroborate such assertions. A logical approach is to compare the risks of the specific product in question with currently acceptable risks. However, the notion of substantive equivalence is not often popular with official bodies because it may expose existing practices as flawed. Yet equivalence is the rational scientific approach (McHughen 2000). As such, an examination of currently acceptable practice for certified forests should reveal comparable requirements for the certification of GM forests.

One of the greatest justifications of anti-GM lobbyists are the potential impacts of GM. Many thus advocate the precautionary principle, at least until the variables are known and a risk assessment can be completed. Yet, consider PEFC prescriptions: "acidification could be controlled by liming. The short term and long term ecological impacts of large scale liming are not fully evaluated" (C5.3, PEFC Sweden, Annex 4, 1999). Liming to combat acidification is considered acceptable, despite its impacts not being fully evaluated. This is contrary to the precautionary principle. Many could draw a parallel between accepted practices such as these and the certification of GM. Indeed, practices considered by some to be very damaging to the environment are justified by increased harvest; "the removal of tops and branches and rotten round wood as wood energy is a supplementary harvest. To compensate for nutrient loss, ashes from wood burning shall be brought back to the forest or compensating fertilising shall be done according to special rules" (Criterion 3:2, PEFC Sweden Annex 5, 1999). Again a value judgement of cost and benefit has been made after assessing potential risks. This would argue that GM applications could be examined on a case by case basis, rather than imposing an outright ban. That is, "focus on the risks of the product, not the process", (McHughen 2000, p.159) just as in conventional forestry.

Since the application of GM cannot be analysed by conventional risk assessment approaches, alternative methods must be carried out. This should be based on scientific fact and substantive equivalence. Because the philosophical bent of individuals is relevant, assessors should be "cognisant and transparent ", combining social awareness with the tenets of rigorous science (Farnham et al. 2000). This ensures a more realistic elucidation and quantification of the trade-offs faced. In order to ensure such deductive distinctions are as objective as possible, knowledge should be integrated through discussion and debate. Nevertheless, it is important to note that "in every study so far, no evidence has been found that GM crops present special risks. The types of risk are exactly the same as for crops modified by the classical plant-breeding methods" (Cook 2000).

If certification systems are to accept GMOs, they have two options. They could accept the judgement of the national systems that currently assess applications for GMO release. This would be very simple, but the differences between national systems could be contentious; some nations are seen as having more rigorous standards than others. However, if certifiers are not satisfied with the stringency of a national system they could impose requirements to meet their own, or another countries, more demanding criteria. Given that it is likely GMO plantations will first appear in developing countries (Owusu 1999), and that national legislation and its implementation may be lacking, it is possible that certification represents the only means of ensuring transgenic releases are safe. As such, it is important for certifiers to recognise that despite never being able to quantify precisely the risks of GM, it is important for them to examine their policies and develop a more sophisticated and realistic approach to the certification of GMOs. Moreover, since it has been shown that silencing of one transgene by another may occur (Matzke & Matzke 1991), widespread monitoring of GMOs has been proposed (Rogers & Parkes 1995). Certification systems could be in a unique position to ensure this monitoring is carried out on a large scale.

As always, it is the application and specifics of the GMO which should be assessed. Risk analysis is very difficult since all variables cannot be known. If GM is to be accepted this will require a flexible certification system. In light of the current permitted practices, from a rational perspective, many applications of GM could be certified, even without risk assessment. Having said this, it is important to note that this is considering GM applications on a case-by-case basis.

CULTURAL ASPECTS OF FORESTS

Forests play a vital role in the life and culture of people around the world. The reverence and adoration of trees has a strong psychological and social foundation in most human cultures. The variety of cultural values and symbolic functions ascribed to forests are as numerous and diverse as the communities and cultures, Forests feature in all aspects of culture: language, history, art, religion, medicine, politics and even social structure.

In many African cultures trees feature in myths and lores. Forest trees serve as link between the sky and earth and is associated with creation as well as the underworld. In parts of West Africa forests provide the venue for many cultural events. The 'arbre a palabre' are locations for social and political meetings. The locations where elders sit under big trees and talk, argue and discuss issues until they agree. It is a place where political, social and judicial decisions are made. In Cote d'Ivoire tree species such as Blighia sapida, Cordia millenii and Bombax buonopozensa are some of the preferred tree species. The Oubangui tribe of Central Africa, plant a tree for every new born child. For female children a fast growing tree species is planted. The child development is linked to the tree growth. If the tree growth declines there is fear for the health of the child and a healer is called upon. When the child is sick she is brought to the tree for ritual treatment. When the tree starts to fruit, the time would have come for the child to marry. When a person dies his/her spirit is believed to go and reside in his/her personal 'birthright' tree.

In other regions of the world also, there exist a relationship between forests and spiritual realm. Budda would sit alone in the depths of the forest in meditation and it was in the midst of a forest that He was shown the four great truths. The Dai people of Yunan province in China believe that the forest is the cradle of human life and that forests are at one with the supernatural realm. The forests in European culture were also considered to be more positive sites of miracles, the search of great spiritual awakenings and the forest itself was held to be form of primitive church or temple. The first temples in Europe were forest groves. Ties to nature manifest themselves most notably in Turkish culture. After conversion to Islam, the importance of trees grew in Turkish culture because Mohammed compared a good Muslim to a palm tree and declared that planting a tree would be accepted as substitute for alms.

SACRED GROVES

In sacred groves are manifested a range of traditions and cultural values of forest throughout the world. Sacred groves are specific forest sites imbued with powers beyond those of human. They are often sites of ancestral burial where people can communicate with their ancestors. Trees within these groves are considered as sacred, housing spirits and providing links to ancestors. In some areas sacred groves are the only intact forests remaining. Although many cultural traditions are disappearing with rapidly changing social and physical environment, sacred groves often remain as valued elements of cultural heritage. The groves are also often the site of ritual healings and location where villagers find particular plant medicine.

Access to most sacred forests are restricted by taboos, codes and custom to particular activities and members of community. Gathering, hunting, woodcutting and farming are strictly prohibited to the Dai people in the holy hills of China. However, control over extractive activities in sacred groves varies in communities and cultures.

In some communities and cultures a complete ban is not in place, and limited collection of fallen wood, fruits from forest floor, medicinal plant collection, honey collection and other activities are permitted, if strictly controlled. Sacred groves in general have survived for many hundred of years and today act as reservoir of biodiversity or library of nature.

In some areas sacred places play a major part in safeguarding critical sites in the hydrological cycle of watershed areas. In different cultures some specific forest resources are revered or serve as religious or cultural symbols. The birch in Scandinavia, larch in Siberia, redwood in California, Fig in India and Iroko in West Africa are widely revered and respected. The oak tree was worshipped by Romans, Druids, Greeks and Celts as the home deities. In Europe fairies were said to have made their homes in old oak trees, departing through holes where branches had fallen; it was considered healing to touch the fairy doors with diseased parts.

Ceiba pentandra (associated with burials and ancestors in the Amazon), Copaifera religiose (associated with fecundity, wealth, power and fame in the South America) and Milicia excelsa (associated with fertility and birth in West Africa) are sacred to the people. Some trees also serve practical judicial roles. Practically they are physical boundary markers that define property and provide evidence of usuary rights in judicial disputes. In many traditions in Senegal and Cote d'Ivoire trees play a central role in the land tenure system. Planted trees evidence land use rights for individual or lineage groups. In Ghana there are court cases that have adjudicated in favour of individuals who have planted and tended naturally regenerating fruit trees on a piece of land for several years without interference as constituting a proof of possession.

TRADITIONAL KNOWLEDGE AND SUSTAINABLE FOREST MANAGEMENT

It is estimated that over 300 million people live in or near forests and obtain part or all of their livelihood and food from forests. Forest is key to non-farm employment for the forest fringe communities. Over the years they have used harvesting methods that are ecologically benign which have ensured their sustenance and survival.

The paradigon of sustainable forest management (SFM) has been widely embraced in national and international policy levels, but it has not yet been implemented to the point where it is appreciably mitigating the negative trends affecting the world forests particularly in the tropics. SFM provides increasingly sophisticated set of policies and tools for managing forests in a more sustainable way. Implementing SFM, however, requires overcoming many of the same economic, political and institutional hurdles that drive deforestation and forest degradation.

Forest management would benefit from incorporation of traditional knowledge of indigenous and local people. Traditional forest related knowledge (TFRK) has long been known to have implications for forest management, conservation of forest biodiversity and identification of forest genetic resources. Traditional knowledge and practices have sustained the livelihoods, cultures and forest resources of local and indigenous communities for centuries. This knowledge most often inter-woven with traditional religious beliefs, customs, folklore, land use practices and community level decision making processes, has historically been dynamic, responding to changing environment, social, economic and political conditions to ensure that forests continue to provide tangible (food, fodder, medicine, water, soil) and non-tangible (spiritual, social, psychological health) benefits to the present and the future generations.

The current limitations of modern science to deal effectively with environmental issues of increasing magnitude and complexity, (e.g. climate change, biodiversity conservation) has opened the door for incorporating other sources of knowledge. Society can learn a great deal from the traditional skills in sustainably managing complex ecological systems. Traditional knowledge holders have developed extensive knowledge about the spatial and temporal distributions of natural resources and behaviours of many natural species and factors that influence them. This knowledge that is distinctive to communities, tribes, cultures arises from personal experience and practices passed on from generation to generation. Over the years local and indigenous people in Africa, Asia and South America have developed a variety of vegetation management practices.

TRADITIONAL KNOWLEDGE AND TRADITIONAL MEDICINE

Traditional medicine refers to health practices, approaches, knowledge and beliefs incorporating plant and animal based medium, spiritual therapies and combination to diagnose, treat and prevent illnesses or maintain well being. In Africa, Asia and Latin America, it is estimated that over 70 percent of the population use traditional medicine to meet their primary health care needs. WHO estimates that in Europe, North America and other industrialized regions over 50 percent of the population have used complementary or alternative medicine at least once and the global market of herbal medicine currently stand over $60 billion and growing steadily. About 25 percent of modern medicines are made from plants and many of these plant based modern medicines are used for the same purposes for which the local and indigenous communities used them. Indeed traditional knowledge underpins the traditional medicine practices.

The construction of traditional knowledge databases and knowledge archives about native groups uses of local plants is a sure way of combating biopiracy, blocking patents by multinational companies and ensuring that holders and users of traditional knowledge on traditional medicinal plants receive fair reward for such knowledge. Database of traditional formulations is needed to allow examiners to compare patent applications with existing traditional knowledge. Indeed, India is one of the countries that has successfully documented traditional knowledge and successfully fought patents granted at the USPTO on turmeric and at EPO on neem (Azadirachta indica). A number of countries have national regulations or in the process of preparing regulations on herb and medicines but the legislative control of medicinal plants has not evolved around structured model. It is proposed that an effective legal system for protection of intellectual property rights of traditional knowledge is established by governments to protect the traditional knowledge of local and indigenous people.

Traditional Knowledge and Climate Change

There are still major gaps in climate science. Traditional knowledge can inform and provide valuable insight into scientific investigations on the effects, impacts and coping strategies for climate change. Local and indigenous people have lived in changing climate over hundreds of years.

As a result of their close relationship and dependence on forests they have developed and use diverse tools to assess the impact of changing climate on their communities and ecosystems and have developed and implemented resilience and adaptive strategies. Local observations of direct effects of climate change corroborates scientific predictions; and include temperature and precipitation changes; coastal erosion; changes in wildlife, pests and water borne diseases distribution; extreme weather events (e.g. drought, flood) and changing weather patterns. In addition many local and indigenous communities are able to forecast impending weather based on early warning signs (typically related to the sky and sea, movements of the sun and moon, changes in plant phenology, and changes in animal behavior). For example there is already a vast store of information at the community level on monsoon prediction. Traditional knowledge can provide scientists and resource managers a long term perspective that is lacking in more typical observations and can establish the relations of historical impacts of land use and climate change. In the context of climate change research local and indigenous communities are providing important source of climate history and baseline data and local scale expertise. Traditional knowledge on climate change does not only augment scientific information but also promote scientific enquiry.

Integration of Traditional Knowledge and Formal Science

Despite their important contribution to SFM and sustainable livelihoods, TFRK and practices are fast disappearing. The negative implication of this loss of TFRK on livelihoods, culture and biodiversity and the capacity of the forest to provide the goods and services remain poorly understood, unappreciated and undervalued by policy makers and the general public in many countries. The question is whether there is any possibility of integrating formal science and ethno-science. Empirical evidence suggests the affirmative. Traditional knowledge systems indeed complement scientific knowledge system by providing practical experience in living within ecosystem and responding to ecosystem change. Traditional knowledge systems and scientific knowledge systems are not exclusive to each other and both are required to achieve SFM. For example a study in Nepal showed that Traditional Knowledge on firewood and fodder values corresponded to scientific assessment. The research revealed that local peoples preference of 16 firewood plants and 23 fodder plants were significantly related to firewood value index (FV1) and fodder value index (FoVi). In vegetation classification and distribution, management and growth characteristics traditional knowledge systems and scientific knowledge systems are in accord with each other.

TREES, WOODS, CULTURE AND IMAGINATION

We have given ourselves a conundrum, without answer. The very notion of creating new forests flies in the face of all we know, and much of what we practice. We know their great age, in the geological sense and domestically. We carved ourselves out of the forest in Britain and in our previous and elsewhere lifetimes. And ever since, we have spent our lives in clearings, on farms, in gardens, in parks, by the wayside, even on buildings, trying to stop trees growing. In parallel we have developed a working relationsliip with trees which has given us great cultural knowledge, affection and wisdom.

We cannot create what nature has spent millions of years learning to do. Forests of a single century demonstrate poverty .... but that is our startiug point, so we must seek and find the ancient and the old. We must protect the 'over mature' and ancient trees, and the nearest we have to ancient woodlands. We need the layers of before, aud if we cannot have it in the trees then it must be revealed in other things.

Beyond these let us relearn about succession ... that much underused branch of ecological insight ... let us watch the course of things. The way that one group of plants and animals will succeed another gradually working to an equilibrium which then stabilises and perpetuates itself dynamically. The gradual supplanting of seral stages moves at a pace which we can appreciate and accommodate. The gentle expansion of those wonderful May trees of the undergrazed upland edge and of greenbelt fields, the old pit tips covered in big trees now but first clothed in scrub. Why have we forgotten how quickly the birch grows and how fast its wood rots ... the perfect pioneer, the beautiful invader.

Trees are as much cultural beings as they are botanical specimens or timber. The tree is a powerful symbolic creature, as well as being a mysterious and miraculous one, hence its use and reuse to help us explain ourselves in the world (The Tree of Knowledge of Good and Evil, the Bo Tree, the Tree of Life). The deciduous tree demonstrates for us the paradox of death and rebirth every year, it embodies longevity, interconnectedness, generosity, hope and also darkness and fear. The forest carries for us many levels of meaning, for example in the boreal fairy stories (Little Red Riding Hood, Hansel and Gretel, The Two Brothers). Bettelheim (1976) in analysing the latter explains:

'The forests, where they go to decide that they want to have a life of their own, symbolizes the place in which inner darkness is confronted and worked through; where uncertainty is resolved about who one is; and where one begins to understand who one wants to be.' Perhaps we have a responsibility like the fairy stories to create for ourselves forests in which we can test ourselves, into which we can read answers which beckon us on to deeper understandings. Forests we can get lost in, safely. Archaic societies have a sophisticated understanding and systems to deal with these things, we flounder around pretending that because they are not economically graspable, not mathematically cageable, not financially fixable that they are not important.

In creating new forests we have to remain sensitive to what already exists, landscapes we may consider bland and those most changed by our activities, carry with them much of significance for those whose life and work has revolved around them. Nowhere should be regarded as a site for rehabilitation, before considered debate with local people. An important leap of the imagination lies in recognising that local knowledge and feeling have much to offer in the process of change. That involvement of local people, as individuals and in groups can offer a rich plural system for planning and energising the forest. Parish Maps focus on the less tangible feelings which people have about their places, starting with the subjective question 'what do you value here?' This makes them the expert, and provides a vehicle for the demonstration of knowledge, emotional attachment and ideas for the future, in a social, practical and creative way.

Trees have been watched, grown, revered, used, eulogised and understood for far longer than the rummaging of scientific arboriculture and the fashions of forestry. We need to know how to interweave new knowledge with the vestiges of the old: the young abstract knowledge, ebullient and often market driven, coining in from the outside looking for a circumstance to practice on, with the old wisdom in place, diffident, severed, unpractised.

In making new forests we must aspire to adding value to the patina of place, conspire to add new layers of significance, to mix in new meaning and detail. Not for the sake of it, but because places without meaning have no long term future and no short term richness. The 21st century forest has to have both. New forests mean new identity for the place, we would assert here the need to strive towards discovering and reinforcing Local Distinctiveness.

Trees have their territories, ecologically defensible but also culturally important. They have helped us to orientate and locate ourselves, trees that are comfortable in their place have added importance to us. D H Lawrence (1928) expressed it in talking of Eastwood, the country of his heart 'To me it seemed, and still seems, an extremely beautiful countryside, just between the red sandstone and the oak trees of Nottinghamshire, and the cold limestone, the ash trees, the stone fences of Derbyshire'; and Oliver Rackham (1986).

'Every oak or alder planted in Cambridge (traditionally a city of willows, ashes, elms and cherry plums) erodes the difference between Cambridge and other places. Part of the value of the native lime tree lies in the meaning embodied in its natural distribution; it is devalued by being made into a universal tree.' Landmark trees remind us of the hard work of the drovers finding their way across country (exotic trees have their place), the parish boundary or the sacred. And ancient trees guarded and venerated, what they are and where they are having melted into reinforcing significance, like the yew tree in the churchyard, or the oak in the centre of the village.

How can the process of creating a new forest 'go with the grain', highlight and accentuate what makes this place different and rich, create novel layers of meaning to celebrate our time? What does the locality already offer, what is here? What do the place names mean, how many of them refer to local practice, trees and woods and other clues to wild nature and our intervetition in the past? What do the field names, names of the woodlands and streams convey? Do you only find cuckoo pounds in Dorset? And what do the new names describe? Where is the history in the landscape (not the past, but history here now), where are the plants and animals particular to this place?

Taking succession seriously demands that our new forests must be created as much by natural regeneration as by planting. Let us see what wants to grow. Let us call it research in the face of global warming if that justification satisfies the fainthearted, but let us learn from nature rather than trying to teach her our short term tricks. Surely we are beyond hypothesis .... it is cheaper, more place sensitive. The challenge is to make the idea exciting to the unimaginative who only see planting by numbers as achievement.

Encouraging natural regeneration outward from existing woodland, from linear hedgerows, from railway lines, road verges and people's gardens is not a passive pursuit. Browsing animals at least will need to be kept at bay.

This implies fences, which also implies gates and stiles. These can all become assets, as visual indicators bearing the marks of locality and passing them on. There may be local patterns of walling or particular ways of laying hedges, there may even be interesting fencing. If not why not work towards the creation of a local pattern? Something which leans on and reinforces the history and the wood use of the place. Stiles and gates too could form part of a simple, sensitive and silent way of letting people know where they are. Extending at whatever pace makes sense in the circumstances, made from local matenals and increasing the demand for them. Made by local people and increasing their involvement and commitment, artists and craftspeople (not necessarily local) can form part of a chain of creativity, helping to reveal, to design, if not make, things which have a richness and meaning born of understanding what is there. A thousand new ideas and as many old ways need to be explored to find ways of passing on interesting knowledge. Postcards, poetry, processions, pruning practice, performance, pollarding, picnics, apprenticeships, apple bobbing. There should be no need for 'interpretation boards', the trees should tell their own stories. Rather than cluttering places with notice boards which are all the same size and design whatever the place and the tale to tell, which reduces rather than enriches both the place and the people. Let's have field names carved in gateposts, seats which gently face the sunset, gates which beckon and give a sense of entering a new dimension.

Let us have cluc after clue about the seasons. Blossom and fruit punctuate the year for us, and nowhere more dramatically and beautifully than through our wild cherry, apple, hawthorn, elder, blackthorn/sloe. Along the roadside and the path, in profusion or in single splendour in the depths of a wood. With care and understanding about the idiosyncrasies of their distribution, we can give other dimensions to the forest. Fruit trees are trees, orchards are groves. Wild and domesticated fruit trees have much to offer to the 21 century forest in town and country.

Orchards have much to offer economically, in wild life and as part of local culture. Community orchards, school orchards, city orchards could offer us pause as well as promise. Old festivals and new, particular like Arbor Day (29 May in Aston on Clun) or countrywide like Apple Day (21 October), are part of the identity of place. With Tree Dressing Day we can draw on traditions from all over the world, to draw in people of many cultures to celebrate the trees they normally take for granted. Seeking permission to perform, to light or to hang things on trees in the public domain during the first weekend in December, demands conversations with tree officers, highway department and police, suddenly bridges are crossed without anxiety, people and institutions are talking and planning together. Overstretched tree officers may begin to share responsibility with the people who 'own' the trees through familiarity. 'Those who arrive at Thekla can see little of the city, beyond the plank fences, the sackcloth screens, the scaffolding ... the wooden catwalks hanging from ropes or supported by the saw horses, the ladders, the trestles. If you ask, 'Why is Thekla's construction taking such a long time?' ..... they answer, 'So that its destruction cannot begin.'

.... 'What meaning does your construction have? ..... What is the plan you are following, the blueprint? They answer 'We will show it to you as soon as the working day is over; we cannot interrupt our work now.' Work stops at sunset. Darkness falls over the building site. The sky is filled with stars. 'There is the blueprint,' they say.' (Italo Calvino, Invisible Cities)

Imagination is crucially needed in the making of our new forests, town or country, for planting trees which seems to be born of goodness, can destroy a place. The creation of new forests has extraordinary potential to raise our sights to help us to reformulate our relations with nature, individually and socially. Our aspiration should be towards something simple and profound. The arts have much to offer. Peter Handke argued 'the duty and potential of all art as the achievement of a glimpse of humanity in the bestiality of human affairs.' Proust illuminates with a description of the poet .....

'(the poet) ... remains before the tree, and tries to close his ears to the sounds from outside and to feel once more what he had felt a moment earlier, when .... the tree had appeared before him, bearing so many white flowers on the tips of its branches that they seemed like innumerable small pellets of snow, as after a thaw. He remains before the tree, but what he seeks is no doubt beyond the tree, for he can no longer feel what he had felt, then all of a sudden he feels it again, but cannot fathom it, can go no further .... it does not seem natural that a poet should remain for an hour before this tree, gazing at how, the spring being imminent, the sure yet unconscious architectural thought known as the double cherry tree genus has arranged these innumerable small fluted white balls which until such time as they wither, will shed a faint perfume through the dark multiplicity of the tree's branches.

The poet gazes and seems to be gazing into himself and into the double cherry tree ..... (Proust, Poetry, or the mysterious laws, letter 1890s) Poetry, like succession and meaning, is about layer upon layer of richness ... literal, implied, concealed, ambiguous, mysterious, paradoxical, dynamic and demanding yet capable of revelation.

We need enchantment, from the most profound moment of realisation that we do not know the magic of the leaf, to the absorbing joy of a tree house. But just as the forestry, recreation and farming professionals need to work with meaning, so do artists. We should demand the best from them, work about and for places and the people who live and spend time there. David Nash with his trees planted with stones, his pruned ash dome and serpentine sycamores. Andy Goldsworthy with his entrance to Hooke Park and Alain Ayers oak leaves for Masham Parish Council both as part of the Common Ground New Milestones project. Jim Partridge with his fence for Grizedale Forest, and his seats for the Woodland Trust with Common Ground. Jamie

MacCullough and his exceptional Beginners Way for the Exmoor Forest. Not a sculpture trail, nothing superficial or contrived, but an adventure which touched people because of the soul searching and honesty which it revealed. This was not about using the forest as a gallery, but the artist's revelation over 18 months hard work, And what of performance, festival, dance, storytelling, poetry.

We must move on from the alienating timber-by-numbers forest, but we do not have to be trapped in some Capability Brown notion of parkland for the leisured classes. We need mixed working woods, gutsy places with real things going on. But that does not mean that the best of the arts cannot be there too. Vitality and authenticity are fired by working, needing, adventuring, imagining, not by the superficial search for fun, for filling in the hours. People need to see the real and to be drawn to reverie. Adding to the depth of significance of the place is a far bigger challenge than planting new forests simply for financial gain. There is so much at stake. We must recognise that it is not an easy matter of economics or ecology, we have even harder tasks to perform which have to do with enchantment, with culture, with reclamation of knowledge of the GOOD, the WILD, the SACRED (Gary Snyder) .....

We have to recapture and progress our long understanding of trees and of the land, to reenfranchise city people and those marooned in over-farmed countryside, to nourishing the imagination. All the economic and ecological knowledge in the world will be of little use to us, if we have no imagination and humanity on which to draw, if we have lost wisdom. 7Common Ground is working with and through the power of ideas, exploring philosophies which intertwine interests in nature, culture and place with poetry. The arts can help us turn our complacent view of ordinary life upside down, can help to bring us back from abstraction, help us to reinvent, to reinterpret and to involve. We must draw in folk who have not the slightest interest in ecologically sustainable futures, who are overwhelmed by the global, bowed by the burden of history, feeling disempowered and disenfranchised by the experts. But starting small and local may help to build what democracy could and should be about. We need to create a culture of wanting to care. Only then shall we know sustainability.

Succession and poetry offer some echoes of each other, the accumulated insights of nature and culture have led both towards richness and complexity, significant layers to be read. We could stop our fight against succession, welcome the trees that want to grow amongst us, plant others of our choosing. Trees at every turn, dreamed into existence by nature and guided by institutions woi'king with local people could help us to reinvent meaning in the places close to us and bring richness back into our everyday lives. We could stop our fight against poetry and recognize that the best of humanity is born of deeper feelings.

Index

A

Abiotic stress 13
Advantages 45, 66, 73, 83, 87, 153, 162, 169, 184, 185, 187, 201, 223, 224, 225, 231, 233, 234, 237, 240, 244, 250, 267
Agriculture 5, 6, 63, 71, 74, 76, 82, 83, 88, 113, 114, 179, 182, 186, 187, 196, 197, 200, 202, 214, 216, 219, 224, 249, 256, 258, 265
Agroforestry 28, 30, 33, 34, 35, 101, 108, 111, 112, 116
Agroforestry systems 33, 34, 35

B

Bioindicator 139, 140
Biology 37, 46, 47, 61, 62, 63, 65, 66, 74, 75, 76, 94, 112, 120, 131, 181, 262, 263
Biotechnologies 74, 130, 131, 168, 178, 181, 183
Biotechnology 2, 6, 9, 74, 76, 81, 86, 87, 91, 95, 128, 129, 130, 131, 168, 173, 174, 175
Bulk density 23

C

Certification 185, 205, 210, 218, 246, 247, 248, 249, 250, 251, 253, 254, 255, 256, 258
Comparisons 38, 55, 108, 138, 263, 272
Complement 6, 42, 47, 123, 133, 233, 276
Consortium 20, 42
Crop improvement 13
Cryopreservation 122, 123, 128, 130, 131, 132, 133, 134, 153, 154, 155, 168, 169

D

Development 8, 11, 12, 13, 15, 16, 24, 38, 39, 42, 45, 46, 47, 54, 61, 62, 64, 66, 67, 68
DNA technology 1

E

Economic 18, 24, 25, 34, 40, 43, 61, 66, 85, 91, 94, 95, 108, 115, 118, 125, 127, 170
Embryogenesis 13, 71, 72, 82, 89, 90, 128, 129, 162, 171, 174, 225, 228
Engineering 9, 11, 12, 13, 67, 71, 74, 75, 76, 81, 84, 92, 124, 125, 129, 130, 168, 170, 171, 173, 174, 200, 206, 269
Environment 2, 4, 5, 9, 11, 12, 15, 18, 20, 21, 22, 23, 24, 25, 26, 31, 32, 33, 37, 42, 46, 47, 61, 62, 63, 65, 68, 85, 88, 90, 91, 94
Evaluating 54, 55, 270
Ex situ conservation 8

F

Farm forestry 34
Food security 33
Forest land 34
Forest plantations 2
Forest Trees 1
Forest trees 1, 5
Forests 2, 5, 18, 19, 20, 23, 24, 25, 26, 35, 36, 46, 56, 57, 61, 62, 71, 91, 93, 114, 117
Fruit production 35
Functions 2, 3, 26, 38, 44, 52, 68, 150, 157, 186, 190, 199, 204, 205, 208, 274

G

Genetic Modification 1
Genetic Transformation 8
Genetic variation 1, 2, 16
Genetically 14, 31, 43, 63, 64, 66, 75, 76, 88, 105, 120, 125, 165, 167, 169, 171, 180
Genomics 13, 37, 40, 42, 44, 45, 46, 47, 62, 66, 67
Germplasm 6, 28, 41, 67, 92, 93, 94, 122, 132, 134, 153, 160, 167, 168
Global system 5

H

Hydroponic 239, 240, 241, 243, 244, 245

I

Improvement 2, 8, 9, 10, 12, 13, 16, 24, 27, 29, 30, 31, 32, 33, 50, 53, 68, 73, 79, 86
Industrial 19, 22, 33, 34, 61, 63, 74, 76, 86, 87, 98, 99, 100, 102, 103, 106, 108, 109
Inoculation 78, 156, 157
Integration 11, 13, 14, 44, 50, 108, 125, 127

J

Juvenility 107, 123, 128, 130, 134, 169, 173, 174, 235, 237, 238

L

Land tenure 35

M

Machining 190, 191, 192, 195
Management 7, 10, 13, 14, 20, 23, 24
Measurement 32, 51, 63, 67, 68, 150, 162, 192
Medicinal 19, 110, 119, 174, 175, 185, 275
Metabolites 41, 81, 86, 90, 155, 174, 175
Methodology 35, 50, 146, 147, 149, 151, 256
Microarray 39, 40, 48, 49, 50, 51, 52, 53, 54
Micropropagation 74
Modification 1, 10, 12, 124, 125, 152, 170
Molecular markers 15, 16, 17

N

Nutritional Improvement 8

O

Organizations 63, 92, 113, 115, 186, 189, 190
Organogenesis 71, 72, 82, 89, 176
Overoptimistic 223

P

Physiological 3, 4, 40, 62, 63, 64, 65, 66, 67, 68, 113, 136, 137, 146, 151, 157, 158, 228, 234, 235, 238, 271
Phytosanitary 244
Plant Genetic Resources 7
Plantations 2, 31, 32
Prediction 2, 49, 50, 51, 52, 53, 54, 55, 56, 57
Propagation 27, 46, 63, 71, 74, 76, 84, 85, 87
Protection 8, 21, 23, 26, 32

R

Recreational 19, 21, 23, 25, 46, 205, 209, 210, 215, 217

S

Sawmilling 190, 191
Scenarios 58, 151, 184, 188
Sequence 9, 15, 16, 17, 19, 23, 37, 38, 39, 42
Silviculture 27
Soil bulk density 23
Soil compaction 22, 23
Soil erosion 2
Soil fertility 34
Somaclonal 76, 89, 125, 157, 161, 162, 163
Somatic 71, 72, 73, 74, 81, 82, 89, 90, 95, 128, 129, 132, 153, 154, 155, 162, 171, 172

T

Techniques 11, 29, 37, 41, 42, 52, 65, 66, 74, 76, 82, 83, 84, 86, 87, 88, 91, 93, 94, 123, 124, 126, 129, 130, 131, 133, 158, 169
Technological 6, 38, 71, 76, 82, 85, 86, 91, 92, 93, 130, 131, 184, 186, 195, 196, 201, 202
Transgene 1, 8, 11, 13, 14, 15, 66, 68, 178, 179, 180, 256, 257, 258, 259, 261, 263, 264, 265, 268, 269, 270, 271, 273
Transgenic 1, 9, 10, 11, 12, 13, 14, 15, 40, 41
Transposon 15, 41
Tree Improvement 30
Tree improvement 29, 30, 31, 32

U

Uniqueness 54, 57
Urban Environment 24